QuickStudy®

for

Geometry

Boca Raton, Florida

DISCLAIMER:

This QuickStudy® Booklet is an outline only, and as such, cannot include every aspect of this subject. Use it as a supplement for course work and textbooks. BarCharts, Inc., its writers and editors are not responsible or liable for the use or misuse of the information contained in this booklet.

©2006 BarCharts, Inc.
ISBN 13: 9781423202578
ISBN 10: 1423202570

BarCharts® and QuickStudy® are registered trademarks of BarCharts, Inc.

Author: Dr. S. B. Kizlik
Publisher:

    BarCharts, Inc.
    6000 Park of Commerce Boulevard, Suite D
    Boca Raton, FL 33487
    www.quickstudy.com

Printed in Thailand

Contents

Study Hints

Note to Student:

Use this QuickStudy® booklet to make the most of your studying time.

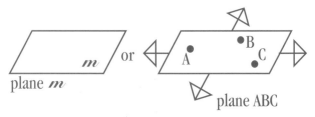

plane *m*   or   plane ABC

All diagrams are set in color, with explanations in red for easy reference. [Note that chapters which deal with only formulas use black text for explanations.]

> **NOTES**
> In coordinate geometry, points can be assigned numbered values, and, as a result, equations of lines can be determined.

QuickStudy® notes provide need-to-know information; read them carefully to better understand key concepts.

Take your learning to the next level with QuickStudy®!

1 — Geometry History

Geometry means "earth measurement." Early peoples used their knowledge of geometry to build roads, temples, pyramids, and irrigation systems.

The more formal study of geometry today is based on an interest in logical reasoning and relationships rather than in measurement alone.

Euclid (300 B.C.) organized Greek geometry into a 13-volume set of books named ***The Elements***, in which the geometric relationships were derived through deductive reasoning. Thus, the formal geometry studied today is often called Euclidean Geometry. This geometry is also called plane geometry because the relationships deal with flat surfaces.

Geometry has undefined terms, defined terms, postulates (assumptions that have not been proven, but have "worked" for thousands of years), and theorems (relationships that have been mathematically and logically proven).

2 Geometric Formulas

Perimeter: The perimeter, P, of a 2-dimensional shape is the sum of all side lengths.

Area: The area, A, of a 2-dimensional shape is the number of square units that can be put in the region enclosed by the sides.

> **NOTES**
> Area is obtained through some combination of multiplying heights and bases, which always form 90º angles with each other; the exception is circles.

Volume: The volume, V, of a 3-dimensional shape is the number of cubic units that can be put in the region enclosed by all the sides.

Square Area:
A=hb; if h=8 and b=8 also, as all sides are equal in a square, then: A=64 square units

Rectangle Area:
A=hb, or A=lw;
if h=4 and b=12 then:
A=(4)(12), A=48 square units

Triangle Area:
A=$^1/_2$ bh; if h=8 and b=12 then:
A=$^1/_2$ (8)(12), A=48 square units

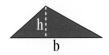

3

Parallelogram Area:
A=hb; if h=6 and b=9 then:
A=(6)(9), A=54 square units

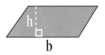

Trapezoid Area:
$A=\frac{1}{2} h(b_1+b_2)$; if h=9, b_1=8 and
b_2=12 then: $A=\frac{1}{2}(9)(8+12)$,
$A=\frac{1}{2}(9)(20)$, A=90 square units

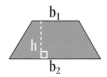

Circle Area:
$A=\pi r^2$; if π=3.14 and r=5 then:
$A=(3.14)(5)^2, A=(3.14)(25)$,
A=78.5 square units

Circumference:
C=2πr, C=(2)(3.14)(5)=31.4 units

Pythagorean Theorem:
If a right triangle has hypotenuse *c*
and sides *a* and *b*, then $c^2=a^2+b^2$:
if a=3 and b=4 then $3^2+4^2=c^2$ and
$25=c^2$ so c=5

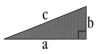

Cube Volume:
$V=e^3$; each edge length, *e*, is equal
to the other edge in a cube,
if e=8 then: V=(8)(8)(8),
V=512 cubic units

Cylinder Volume:
$V=\pi r^2 h$;
if radius r=9 and h=8 then:
$V=\pi(9)^2(8), V=3.14(81)(8)$,
V=2034.72 cubic units

Rectangular Prism Volume:
V=lwh; if l=12, w=3 and h=4 then:
V=(12)(3)(4), V=144 cubic units

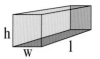

Cone Volume:
$V=\frac{1}{3}\pi r^2 h$; if r=6 and h=8 then:
$V=\frac{1}{3}\pi(6)^2(8)$,
$V=\frac{1}{3}(3.14)(36)(8)$,
V=301.44 cubic units

Triangular Prism Volume:
V=(area of triangle)h;

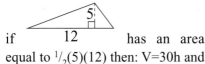

if   has an area
equal to $\frac{1}{2}(5)(12)$ then: V=30h and
if h=8 then: V=(30)(8), V=240
cubic units

Rectangular Pyramid Volume:
$V=\frac{1}{3}$(area of rectangle)h; if l=5 and
w=4 the rectangle has an area of 20,
then: $V=\frac{1}{3}$(20)h and if h=9 then:
$V=\frac{1}{3}$(20)(9), V=60 cubic units

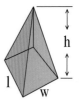

Sphere Volume:
$V=\frac{4\pi r^3}{3}$; if radius r=5 then:

$V = \frac{4(3.14)(5)^3}{3}, V = \frac{1570}{3},$

523.3 cubic units

3 Descriptions of Undefined Terms

Point

■ A **point** may be described as a location with no length, no width, and no depth.

■ A point is always named with a capital letter; it is usually located by using a dot about the size of a period, although true points cannot really be drawn because they have no dimensions.
Ex: .A (indicates point A).

Line

■ A **line** may be described as a set of points going straight on forever in two opposite directions; lines are straight and never end; there is never a need to use the phrase "straight line" because lines are straight; if something is not straight it cannot be a line, but might be a curve instead; lines have length, but no width and no depth; lines cannot be truly drawn because they lack width and depth; representations of lines are drawn with arrows at each end indicating that the line has no end.

■ Lines are usually named in one of two ways; the line containing points K and M may be named by either:

◆ Using any 2 (never more than 2) points on the line with a line indicator above the points, for example $\overleftrightarrow{KM}$ or $\overleftrightarrow{MK}$ (the order of the points doesn't matter); the line indicator above the capital letters always points horizontally from side to side and never any other direction; it is the actual location of the points in space that determines the location and direction of the real line, not the direction of the line indicator above the capital letters in the notation; or

◆ By using the lower case script letter l with a number subscript, for example: l_1 or l_2.

NOTES
In coordinate geometry, points can be assigned numbered values, and, as a result, equations of lines can be determined.

Planes

■ A **plane** may be described as a set of points going on forever in all directions, except any direction that creates depth; imagine the very surface of a perfectly flat piece of paper extending on forever in every direction but having no thickness at all; the result would be a situation such that when any 2 points in this plane are connected by a line, all points in the line are also in the plane; planes have length and width, but no depth.

■ Planes are simply referred to as "plane *m*" or "plane ABC (any 3 points on the plane that are not on the same line)" or "the plane containing —— (whatever pertains to the discussion)"; planes cannot be drawn; representations of planes are usually drawn as parallelograms, either with the arrows indicating that the points go on forever or without the arrows even though the points do go on forever.

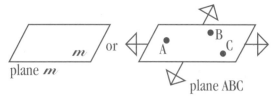

plane *m*

plane ABC

Postulates

Postulates are statements that have been used and accepted for centuries without formal proof; these are postulates:

■ A line contains at least 2 points, and any 2 points locate exactly 1 line.

■ A plane contains at least 3 points that are not all on the same line, and any 3 points that are not on the same line locate exactly 1 plane; therefore, a line and 1 point not on the line also locate exactly 1 plane.

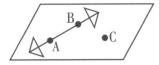

■ Any 3 points locate at least 1 plane.

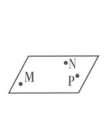   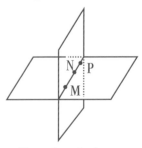

1 plane
when the 3 points are
not on the same line

More than 1 plane
when the 3 points are
on the same line

■ If 2 points of a line are in a plane, then the line is in the plane; or, if 2 points are in a plane, then the line containing the 2 points is also in the plane.

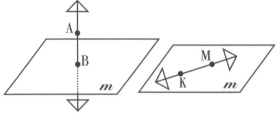

$\overleftrightarrow{AB}$ intersects plane *m*, but is not contained in plane *m*

$\overleftrightarrow{KM}$ is contained in plane *m*, because both points K and M are in plane *m*

■ If 2 planes intersect, then their intersection is a line.

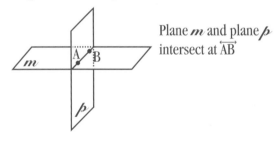

Plane *m* and plane *p* intersect at $\overleftrightarrow{AB}$

Defined Terms

There are many defined terms of plane geometry; the definitions of these terms will be given by topic groups throughout the study booklet rather than grouped in one big list.

NOTES

Postulates (or axioms) are relationships and statements that have worked for centuries and are accepted without mathematical proof; **theorems** are relationships and statements that have been proven mathematically; postulates and theorems are given throughout this guide; they also are stated under the various topics of geometry; sometimes they are labeled as postulates or theorems, and sometimes they are simply stated and not labeled.

Space

Refers to the **set of all points;** space goes on forever in every direction, and therefore has length, width and depth; space has no special notations; it is simply referred to as space; space contains at least 4 points that are not all on the same plane.

General Terms

■ ≅ or **congruent** shapes are the same shape and size; therefore, after some movement of the shapes they can be made to fit exactly on top of one another.

> **NOTES**
> Corresponding parts of congruent polygons are congruent; that is, once the polygons have been moved around to match up perfectly, then the parts that match (correspond) are congruent.

Consult angle and line notation for these examples:

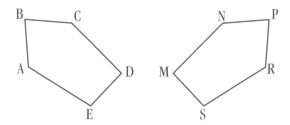

Pentagon ABCDE ≅ Pentagon RPNMS such that matching sides have equal lengths and matching angles have equal measures. If side $\overline{AB}$=2 feet then side $\overline{PR}$=2 feet. Corresponding parts are: $\overline{AB}\cong\overline{PR}$; $\overline{BC}\cong\overline{NP}$; $\overline{CD}\cong\overline{MN}$; $\overline{DE}\cong\overline{MS}$; $\overline{EA}\cong\overline{SR}$; ∢A≅∢R; ∢B≅∢P; ∢C≅∢N; ∢D≅∢M; ∢E≅∢S;

■ ~ or **similar** shapes are the same shape, but can be different sizes; thus, congruent shapes are also similar shapes, but similar shapes are not necessarily congruent shapes.

NOTES

2 or more similar polygons have corresponding (matching) interior angles of 1 polygon congruent to the corresponding interior angles of the other, but the corresponding sides are proportional, not necessarily congruent.

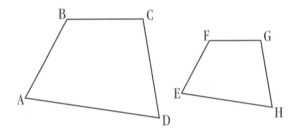

Quadrilateral ABCD ~ Quadrilateral EFGH; therefore matching angles have equal measures and matching sides have proportional measures; For example: If AB = 8 and EF = 4 and BC = 7 then FG = 3.5; that is, $\frac{AB}{EF} = \frac{BC}{FG} = \frac{CD}{GH} = \frac{DA}{HE}$ and $\angle A \cong \angle E$, $\angle B \cong \angle F$, $\angle C \cong \angle G$ and $\angle D \cong \angle H$.

■ = or **equal** can apply to sets of points being exactly the same set or to numerical measurements being exactly the same number values.

■ ∪ or **union** refers to putting all of the points together and describing the result.

■ ∩ or **intersection** refers to describing only those points that are common to all sets involved in the intersection or to describing the points where indicated shapes touch.

Lines

While the word **line** has no formal definition, lines are always straight and named. Also there are some particular terms that refer to relationships involving lines:

■ **Collinear** points are points that are on the same line.

Points A, B and C are
collinear

Points D, E and F are
noncollinear

■ **Noncollinear** points are points that are not on the same line; any 3 noncollinear points are on some plane, however, and so are **coplanar**.

Points M, N and P are
noncollinear, but are
on the same plane,
and so are coplanar

■ **Categories of Lines**
 ◆ **Intersecting Lines**
  • Intersecting lines share 1 and only 1 point in common.

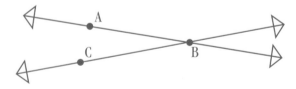

$\overleftrightarrow{AB} \cap \overleftrightarrow{CB} = B$; the 2 lines intersect at point B

• Intersecting lines lie in 1 plane.

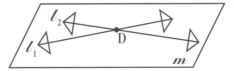

ℓ_1 and ℓ_2 intersect at point D and lie in plane m

◆ Perpendicular Lines

• Perpendicular lines are lines that intersect and form 90° angles (see the section on angles) at the intersection; the 90° angles are indicated on diagrams by drawing a small square in the corner by the vertex of the angle.

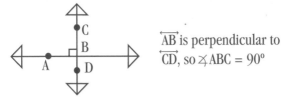

$\overrightarrow{AB}$ is perpendicular to $\overleftrightarrow{CD}$, so $\angle ABC = 90°$

• Through a point not on a line exactly 1 perpendicular can be drawn to the line.

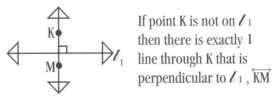

If point K is not on ℓ_1 then there is exactly 1 line through K that is perpendicular to ℓ_1, $\overleftrightarrow{KM}$

• ⊥ means "is perpendicular to"; therefore, $\ell_1 \perp \ell_2$ is read "line 1 is perpendicular to line 2."

Theorem: The shortest distance from any point to a line or to a plane is the perpendicular distance.

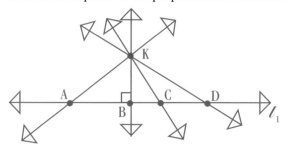

The shortest distance from point K to ℓ_1 is the distance from K to B

◆ **Transversal Lines**
  • A transversal is a line that intersects 2 or more coplanar lines at different points.

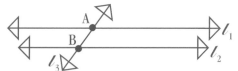

ℓ_3 is a transversal because it intersects ℓ_1 at A and ℓ_2 at B

◆ **Parallel Lines**
  • Parallel lines lie in the same plane (coplanar) and share no points in common; i.e., they do not intersect.

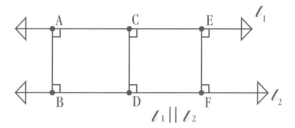

- Parallel lines go in the same directions and never touch; parallel lines are everywhere the same distance apart.
- Through a point not on a line exactly 1 parallel can be drawn to the line.
- $||$ means "is parallel to," so $\ell_1 || \ell_2$ is read "line 1 is parallel to line 2."

Theorem: If 3 or more parallel lines cut off equal segments on 1 transversal, then they cut off equal segments on every transversal that they share.

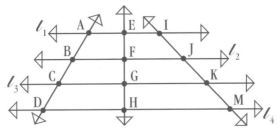

If $\ell_1 || \ell_2 || \ell_3 || \ell_4$ and if AB = BC = CD, then EF = FG = GH and IJ = JK = KM

NOTES
Special angles that result when 2 or more lines are intersected by a transversal are discussed under the topic of angles.

◆ **Skew Lines**

• Skew lines are not in the same plane (noncoplanar) and never touch; they go in different directions.

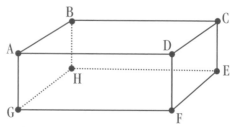

$\overleftrightarrow{AB}$ and $\overleftrightarrow{DC}$ are parallel, but $\overleftrightarrow{AB}$ and $\overleftrightarrow{GF}$ are skew because they never touch and they go in different directions

Planes

While the word **"plane"** has no formal definition, planes are flat with length and width, and planes are named. Also, note definitions for the following terms:

■ **Coplanar** means in the same plane; therefore, coplanar points lie in the same plane.

■ **Noncoplanar** means not in the same plane; 3 points cannot be noncoplanar because there is some plane that exists that contains any 3 points; the smallest number of points that can be noncoplanar is 4.

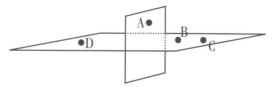

Points B, C and D are coplanar, but points B, C, D and A are noncoplanar

■ A line and a plane are parallel if they do not touch or intersect.

■ 2 or more planes are parallel if they do not touch or intersect.

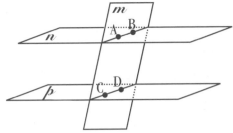

Planes *n* and *p* are parallel;
Planes *m* and *n* intersect at $\overleftrightarrow{AB}$
Planes *m* and *p* intersect at $\overleftrightarrow{CD}$
Note: $\overleftrightarrow{AB} \parallel \overleftrightarrow{CD}$ because plane *n* is parallel to plane *p*

Theorem: If 2 parallel planes are both intersected by a third plane, then the lines of intersection are parallel.

Line Segments

■ A **line segment** is the set of any 2 points on a line (the endpoints) and all the collinear points between them; a line segment is named using the 2 endpoints and a bar notation drawn above these two points; for example, $\overline{PR}$ includes endpoints P and R and all of the collinear points between them.

NOTES

$\overline{PR} = \overline{RP}$ because both notations name exactly the same set of points.

■ The **union** of 2 line segments depends on the location of each; for example, these conditions could exist:

◆ They do not touch: Then $\overline{AB} \cup \overline{CD} =$ {all points on $\overline{AB}$ together with all the points on $\overline{CD}$}.

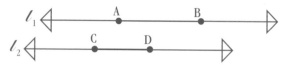

◆ They touch in 1 point: Then $\overline{EF} \cup \overline{HI} =$ {all points on $\overline{EF}$ together with all of the points on $\overline{HI}$}.

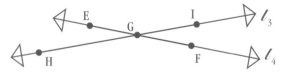

◆ They touch in more than one point: Then

$\overline{AB} \cup \overline{CD} = \overline{AB}$ when $\overline{CD}$ $\overline{EF} \cup \overline{GH} = \overline{EH}$ when $\overline{EF}$
is contained in $\overline{AB}$ and $\overline{GH}$ overlap

■ The **intersection** of two line segments is either no points (they do not touch), 1 point, or another line segment; for example: these conditions could exist.

◆ They do not touch: Then $\overline{AB} \cap \overline{CD}$ = the empty set because they have no points in common. $\overline{AB} \cap \overline{CD} = \emptyset$

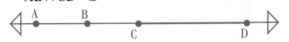

◆ They touch in 1 point: Then $\overline{EF} \cap \overline{FG} = \{\text{point } F\}$.

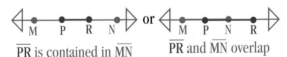

◆ They touch in more than 1 point:
Then $\overline{MN} \cap \overline{PR} = \overline{PR}$ or $\overline{MN} \cap \overline{PR} = \overline{PN}$

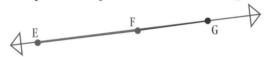

$\overline{PR}$ is contained in $\overline{MN}$ $\overline{PR}$ and $\overline{MN}$ overlap

■ The length of a line segment or the **distance** between 2 points is a numerical value; the notation for distance is 2 capital letters with no bars or additional notations above the letters; for example: The distance between point T and point S is indicated by the notation TS with no commas and no additional notations above the 2 capital letters.

$\overline{TS}$ means line segment with endpoints T and S.

TS means the length of $\overline{TS}$.

■ The **midpoint** of a line segment is a point exactly in the middle of the 2 endpoints; for example: Point R is the midpoint of $\overline{TS}$ if point R is on $\overline{TS}$ and TR = RS; also notice that TR + RS = TS.

TR = RS so R is the midpoint of $\overline{TS}$

■ The **bisector** of a line segment intersects the line segment at its midpoint; a bisector can be a point, line, line segment, ray, plane, as well as other shapes.

■ The **perpendicular bisector** of a line segment intersects the line segment at its midpoint and forms 90° angles at the intersection (see the section on angles); a square in the corner by the vertex of the angle indicates a 90° angle.

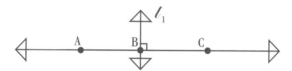

ℓ_1 is the $\perp$ bisector of $\overline{AC}$ because it forms 90° angles at the midpoint, B, of $\overline{AC}$

Theorem: If a point lies on the perpendicular bisector of a line segment, then the point is equidistant (equal distances) from the endpoints of the line segment.

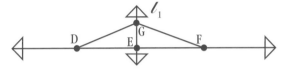

ℓ_1 is the $\perp$ bisector of $\overline{DF}$ so GD = GF, the distance from G to D and G to F are equal

Theorem: If a point is equidistant from the endpoints of a line segment, then the point lies on the perpendicular bisector of the line segment.

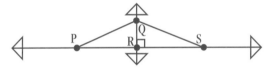

If PQ = QS and PR = RS, then $\overleftrightarrow{QR}$ must be the $\perp$ bisector of $\overline{PS}$

■ To **trisect** a line segment, separate it into 3 other line segments such that the sum of the lengths of the 3 segments is equal to the length of the original line segment; for example: $\overline{AB}$ has been trisected at points C and D because AC + CD + DB = AB.

Rays

■ A **ray** is the set of collinear points going in 1 direction from a point (the endpoint of the ray) on a line; the length of a ray cannot be measured because it has only 1 endpoint, the notation for writing a ray is 2 capital letters indicating first, the endpoint of the ray, and second, any other point on the ray; a bar with an arrowhead on the right end must be drawn above the letters to indicate that it is a ray; for example: $\overrightarrow{AB}$ has the endpoint A and goes forever in the direction of point B; however, $\overrightarrow{BA}$ has the endpoint B and goes on forever in the direction of point A; notice that $\overrightarrow{AB}$ and $\overrightarrow{BA}$ do not contain the same set of points, therefore $\overrightarrow{AB} \neq \overrightarrow{BA}$.

■ The union or intersection of 2 rays depends on their relative positions; for example:

◆ If the 2 rays do not touch, the union is simply all the points on both rays, and there is no intersection or common points.

◆ If the 2 rays touch in 1 and only 1 point, but not at the endpoint, then the union is all the points on both rays, and the intersection is that 1 point where they touch.

◆ If the 2 rays touch in 1 and only 1 point, the endpoint, then **the union is an angle,** and the intersection is the endpoint.

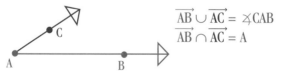

$$\overrightarrow{AB} \cup \overrightarrow{AC} = \angle CAB$$
$$\overrightarrow{AB} \cap \overrightarrow{AC} = A$$

◆ If the 2 rays touch in more than 1 point then the union is a line, and the intersection is a line segment,

$$\overrightarrow{AB} \cup \overrightarrow{BA} = \overleftrightarrow{AB}$$
$$\overrightarrow{AB} \cap \overrightarrow{BA} = \overline{AB}$$

or the union is a ray and the intersection is another ray.

$$\overrightarrow{KM} \cup \overrightarrow{PM} = \overrightarrow{KM}$$
$$\overrightarrow{KM} \cap \overrightarrow{PM} = \overrightarrow{PM}$$

■ **Opposite rays** are collinear rays that share only a common endpoint and go in opposite directions.

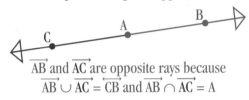

$\overrightarrow{AB}$ and $\overrightarrow{AC}$ are opposite rays because
$\overrightarrow{AB} \cup \overrightarrow{AC} = \overleftrightarrow{CB}$ and $\overrightarrow{AB} \cap \overrightarrow{AC} = A$

Angles

■ An **angle** is the union of 2 rays that share 1 and only 1 point, the endpoint of the rays; the sides of the angle are the rays and the vertex of the angle is the common endpoint of the rays; the interior of the angle is all the points between the 2 sides of the angle; the plural of vertex is vertices.

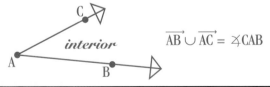

interior

$$\overrightarrow{AB} \cup \overrightarrow{AC} = \angle CAB$$

NOTES
Notice that the symbols are both flat on the bottom, unlike the "less than" symbol <

An angle may be named using the vertex only if there is only 1 angle at the vertex; if more than 1 angle is present at the vertex then the angle must be named by either using 3 points of the angle, with the vertex listed as the middle letter, or by assigning the angle a numerical name in the interior of the angle, close to the vertex.

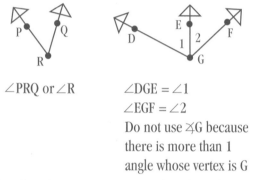

∠PRQ or ∠R

∠DGE = ∠1

∠EGF = ∠2

Do not use ∠G because there is more than 1 angle whose vertex is G

■ **Overlapping angles** are angles that share some common interior points.

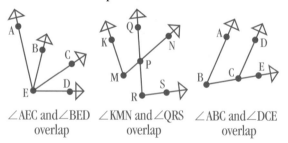

∠AEC and ∠BED overlap

∠KMN and ∠QRS overlap

∠ABC and ∠DCE overlap

■ Angles are **measured** using a protractor and degree measurements; there are 360° in a circle; placing the center of a protractor at the vertex of an angle and counting the degree measure is like putting the vertex of the angle at the center of a circle and comparing the angle measure to some of the degrees of the circle.

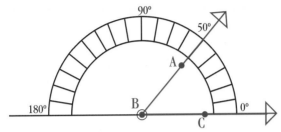

The measure of ∠ABC = m∠ABC = 50°
Notice: m∠ABC means the measure of the
angle in degrees

■ An **acute** angle is an angle that measures less than 90°.

■ An **obtuse** angle is an angle that measures more than 90°.

■ A **right** angle is an angle that measures exactly 90°; it is indicated on diagrams by drawing a square in the corner by the vertex of the angle.

■ A **straight** angle is an angle that measures exactly 180°.

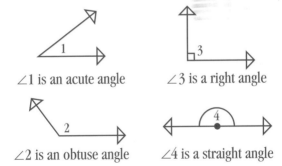

∠1 is an acute angle   ∠3 is a right angle

∠2 is an obtuse angle   ∠4 is a straight angle

■ **Complementary angles** are 2 angles whose measures total 90º.

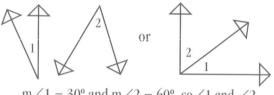

m∠1 = 30º and m∠2 = 60º, so∠1 and ∠2 are complementary angles

Theorem: If 2 angles are complements of the same angle then they are equal in measure.

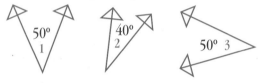

∠1 and ∠2 are complements, ∠2 and ∠3 are complements, therefore, m∠1 = m∠3; = measurements

Theorem: If 2 angles are complements of congruent angles (angles having the same degree measures), then they are congruent.

■ **Supplementary angles** are 2 angles whose measures total 180º.

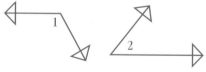

m∠1 = 120º and m∠2 = 60º, so ∠1 and ∠2 are supplementary angles because m∠1 + m∠2 = 180º

Theorem: If 2 angles are supplements of the same angle, then they are congruent (have the same degree measures).

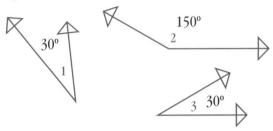

∠1 and ∠2 are supplements; ∠2 and ∠3 are supplements, therefore, m∠1 = m∠3; ∠1 ≅ ∠3

Theorem: If 2 angles are supplements of congruent angles, then they are congruent.

■ **Vertical angles** are 2 angles that share only a common vertex and whose sides form lines.

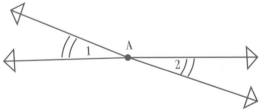

∠1 and ∠2 are vertical angles; they share the same vertex, A, and their sides form lines; m∠1 = m∠2

NOTES
Angles with equal measures are indicated by equal number of curves in the angle interiors.

Theorem: Vertical angles are congruent and have equal measures.

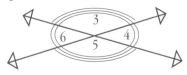

∠3 and ∠5 are vertical angles; m∠3 = m∠5
∠6 and ∠4 are vertical angles; m∠6 = m∠4
∠4 and ∠5 are supplements
∠5 and ∠6 are supplements
∠6 and ∠3 are supplements
∠3 and ∠4 are supplements

■ **Adjacent angles** are 2 angles that share exactly 1 vertex and 1 side, but no common interior points; i.e., they do not overlap.

∠ABC and ∠CBD are adjacent ∠s
∠ABC ∩ ∠CBD = $\overrightarrow{BC}$
But ∠ABD and ∠CBD are not adjacent angles because they share common interior points and overlap

■ An angle is **bisected** by a ray or a line that contains the vertex of the angle, is in the interior of the angle, and separates the angle into 2 adjacent angles with equal measures.

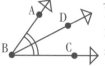

$\overrightarrow{BD}$ bisects ∠ABC because m∠ABD = m∠DBC and ∠ABD and ∠DBC are adjacent

Theorem: If a point lies on the bisector of an angle, then the point is equidistant (equal distances) from the sides of the angle.

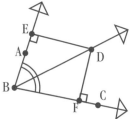

ED = FD because D is on $\overrightarrow{BD}$ which bisects ∠ABC

NOTES
Distance from a point to a line is always measured on the perpendicular line segment that connects the point and the line.

Theorem: If a point is equidistant (equal distances) from the sides of an angle, then the point lies on the bisector of the angle.

■ An angle is **trisected** by rays or lines that contain the vertex of the angle and separate the angle into 3 adjacent angles (in pairs) that all have equal measures.

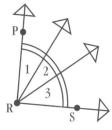

∠PRS is trisected because m∠1 = m∠2 = m∠3, and angles 1, 2 and 3 do not overlap

■ Angles formed when 2 or more lines are intersected by a transversal:

◆ **Interior angles** are formed with the rays from the 2 lines and the transversal such that the interior regions of the angles are between the 2 lines.

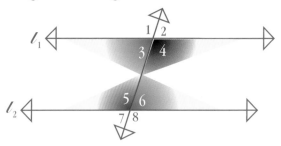

Angles 3, 4, 5 and 6 are interior angles
Angles 1, 2, 7 and 8 are exterior angles

• **Alternate interior angles** are 2 interior angles that have different vertices and are on opposite sides of the transversal.

Theorem: If the lines are parallel, then the alternate interior angles are equal in measure, and if the alternate interior angles are equal in measure, then the lines are parallel.

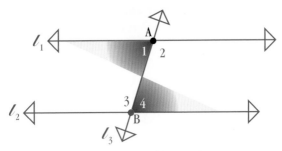

Angles 1, 2, 3 and 4 are interior angles;
Angles 1 and 4 are alternate interior angles
because they have different vertices and are on
opposite sides of the transversal l_3; $\angle 1$ has vertex
A and its interior is on the left of l_3; $\angle 4$ has
vertex B and its interior is on the right side of l_3;
Additionally $\angle 2$ and $\angle 3$ are alternate interior
angles. If l_1 and l_2 are parallel, then $\angle 1 = \angle 4$
and $\angle 2 = \angle 3$

- **Same side interior angles** are 2 interior angles that have different vertices and are on the same side of the transversal.

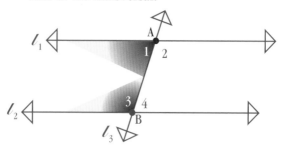

$\angle 1$ and $\angle 3$ are same side interior angles; they are both on the left side of l_3 and $\angle 1$ has vertex A while $\angle 3$ has vertex B; $\angle 2$ and $\angle 4$ are also same side interior angles

Theorem: If the lines are parallel, then the same side interior angles are **supplementary** (total 180°), and if the same side interior angles are supplementary, then the lines are parallel. If $l_1 \| l_2$ above then $m\angle 1 + m\angle 3 = 180°$ and $m\angle 2 + m\angle 4 = 180°$

◆ **Exterior angles** are formed when 2 or more lines
are intersected by a transversal; they are formed
by the lines and the transversal such that the
interior regions of the angles are not between the
2 lines, but are outside and away from the 2 lines.

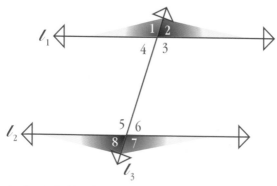

Angles 1, 2, 7 and 8 are exterior angles; $\angle 1$ and $\angle 7$
are alternate exterior angles, as are $\angle 2$ and $\angle 8$

• **Alternate exterior angles** are exterior angles
that have different vertices and are on opposite
sides of the transversal; if the lines are parallel,
then the alternate exterior angles are equal in
measure, and if the alternate exterior angles are
equal in measure, then the lines are parallel.
If ℓ_1 and ℓ_2 above are parallel then $m\angle 1 = m\angle 7$
and $m\angle 2 = m\angle 8$

◆ **Corresponding angles** are angles that have different vertices, are on the same side of the transversal, and are in the same positions relative to the lines and the transversal; one of the pair of corresponding angles is an interior angle and the other is an exterior angle.

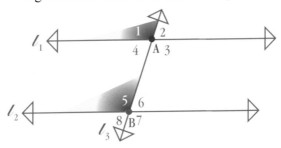

∠1 and ∠5 are corresponding angles; the interior of ∠1 is on the left of l_3 on top of l_1, ∠1 has vertex A while ∠5 has vertex B; the interior of ∠5 is also on the left of l_3 and on top of l_2; if you slide the 4 angles at vertex B up the transversal, l_3, to vertex A, then ∠5 would land on ∠1, ∠6 on ∠2, ∠7 on ∠3, and ∠8 on ∠4; so these are all pairs of corresponding angles:

∠1 and ∠5
∠2 and ∠6
∠3 and ∠7
∠4 and ∠8

If $l_1 || l_2$, then these corresponding angles are equal in measure; thus:

m∠1 = m∠5
m∠2 = m∠6
m∠3 = m∠7
m∠4 = m∠8

Postulate: If the lines are parallel, then the corresponding angles are equal in measure, and if the corresponding angles are equal in measure, then the lines are parallel.

◆ **Right angles** (90°) are formed when a transversal is perpendicular to the lines that it intersects.

Theorem: If a transversal is perpendicular to 1 of 2 parallel lines, then it is also perpendicular to the other.

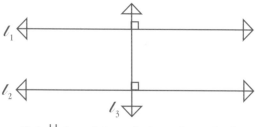

If $\ell_1 \,||\, \ell_2$ and $\ell_3 \perp \ell_1$ then $\ell_3 \perp \ell_2$ also

Polygons

■ **Polygons** are plane shapes that are formed by line segments that intersect only at the endpoints. These intersecting line segments create one and only one enclosed interior region.

Polygon ABCDE

Not a polygon; this shape is not closed

Not a polygon;
One side is not
a line segment

2 polygons
intersecting
at point A

■ Polygons are named by listing the endpoints of the line segments in order going either clockwise or counterclockwise, starting at any one of the endpoints.

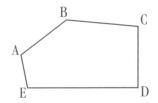

This may be called polygon ABCDE or polygon CBAED or ...

■ The **sides** of polygons are line segments; polygons are all of the points on the sides (line segments) and vertices.

The sides of the polygon shown above are $\overline{AB}$, $\overline{BC}$, $\overline{CD}$, $\overline{DE}$ and $\overline{EA}$. The vertices are points A, B, C, D and E.

■ The **vertices** (or vertexes) of polygons are the endpoints of the line segments.

■ **Diagonals** of a polygon are line segments whose endpoints are vertices of the polygon, but diagonals are not line segments that are the sides of the polygons.

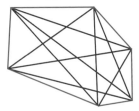

The red line segments are sides of the polygon; The blue line segments are the diagonals

■ The **interior** of a polygon is all of the points in the region enclosed by the sides. The exterior of a polygon is all of the points on the plane of the polygon, but not on the sides nor in the interior of the polygon.

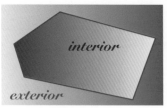

The points indicated in red are the points of the polygon

■ The **interior angles** of a polygon are the angles that have the same vertex as one of the vertices of the polygon and have sides and interiors that are also sides and interiors of the polygon. Every polygon has as many interior angles as it has vertices.

Angles 1, 2, 3 and 4 are interior angles of the polygon

■ **Concave** polygons have at least one interior angle whose measure is more than 180°.

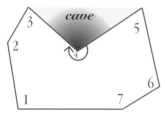

Interior angle 4 is more than 180° so this polygon is concave; Notice the "cave" in concave and in the concave polygon

■ **Convex** polygons have no interior angles that are more than 180º. All interior angles have measures that are each less than 180º.

Convex polygons have no "caves"

Theorem: The sum of the measures of the interior angles of a convex polygon with **n** sides is (**n** - 2) 180º.

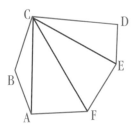

Pick any one vertex and draw all diagonals from that one vertex; There will always be 2 less triangles than the number of sides of the polygon; Since each triangle has angles that total 180º, multiply the number of triangles by 180º, thus the formula $(n-2)180°$; In polygon ABCDEF, $n=6$ because there are 6 sides; Using C and drawing all diagonals from point C creates 4 triangles so $(n-2)180° = (6-2)180° = (4)180° = 720°$

NOTES

To find the measures of each interior angle of a regular polygon, find the sum of all of the interior angles and divide by the number of interior angles. Thus the formula: $\dfrac{(n-2)180°}{n}$

If hexagon ABCDEF, above, were a regular hexagon all angles would be equal, so

$$\frac{(n-2)180°}{n} = \frac{720°}{6} = 120° \text{ each.}$$

■ **Exterior angles** of polygons are formed when the sides of the polygon are extended. Each exterior angle has a vertex and one side which are also a vertex and one side of the polygon. The second side of the exterior angle is the extension of the polygon sides.

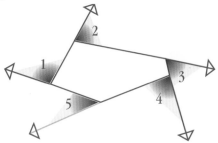

Angles 1, 2, 3, 4 and 5 are exterior angles of the polygon, and their sum equals 360°

Theorem: The sum of the measures of the exterior angles of any convex polygon, using one exterior angle at each vertex, is 360°.

■ **Regular polygons** are polygons with all side lengths equal and all interior angle measures equal.

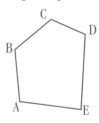 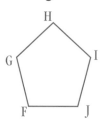

Polygon ABCDE   Regular polygon FGHIJ because
FG = GH = HI = IJ = JF and
$m\angle F = m\angle G = m\angle H = m\angle I = m\angle J$

Classifications of Polygons

■ Polygons are classified by the number of sides, which is equal to the number of vertices.

■ The side lengths are not necessarily equal unless the word "regular" is also used to name the polygon. A regular polygon has equal side lengths and equal interior angle measurements.

■ Categories

Polygon Name	Number of Sides
Triangles	3
Quadrilaterals	4
Pentagons	5
Hexagons	6
Heptagons (Septagons)	7
Octagons	8
Nonagons	9
Decagons	10
n-gons	*n*

Special Polygons

■ **Triangles** are polygons with 3 sides. Each triangle also has 3 vertices. The symbol for triangle is Δ.

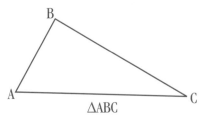

ΔABC

◆ Altitudes and bases of triangles are used to find the areas:

- The **altitude (height)** of a triangle is the line segment whose endpoints are a vertex of the triangle and the point on the line containing the opposite side of the triangle where a 90° angle is formed. The altitude is perpendicular to the line containing the side opposite the vertex of the triangle. Since a triangle has 3 vertices, every triangle has 3 altitudes.

- The **base** of a triangle is the side of the triangle that is on the line that is perpendicular to the altitude. Since every triangle has 3 altitudes, every triangle has 3 bases. Each altitude has a different side that is the base.

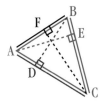

In ΔABC, $\overline{BD}$ is the altitude for base $\overline{AC}$, $\overline{AE}$ is the altitude for base $\overline{BC}$, and $\overline{CF}$ is the altitude for $\overline{AB}$ Notice that each altitude forms a 90° angle with its base

- The **area of a triangle** may be found by applying this formula: $A = \frac{1}{2}ab$ where

  A   means area

  a   means altitude

  b   means base. Any base and its altitude may be used.

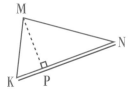

In $\triangle$KMN, if base KN=18 and altitude MP=12 then $A = \frac{1}{2} \bullet 12 \bullet 18 = 108$ units^2

◆ Triangles are **classified** in 2 ways, by side lengths and by angle measurements.

  - When classified by **side lengths**, a triangle is either:

    › **scalene**; that is, no side lengths are equal,

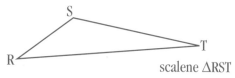

scalene $\triangle$RST

    › **isosceles**; that is, at least 2 side lengths are equal,

Isosceles $\triangle$XYZ because YZ=ZX
Notice: The marks on the sides indicate lengths are equal

> **equilateral**; that is, all 3 side lengths are equal.

△ABC is equalateral because
AB=BC=CA;
△ABC is also isosceles because
at least 2 sides are equal

NOTES
An equilateral triangle is also an isosceles triangle.

- When classified by **angle measurements**, a triangle is either:
  > **obtuse**; that is, one angle measurement is more than 90°,

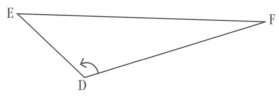

△DEF is obtuse because $m\angle D > 90°$;
Notice that both $m\angle E < 90°$ and $m\angle F < 90°$

  > **right**; that is, one angle measurement is equal to 90°,

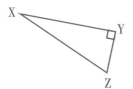

△XYZ is a right triangle
because $m\angle Y = 90°$

> **acute**; that is, all 3 angle measurements are less than 90º.

NOTES

If all 3 angles are equal, then the triangle is called equiangular.

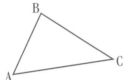

ΔABC is an acute triangle

NOTES

Triangles are classified using one side classification name and one angle classification name; therefore, a triangle classification uses two words.

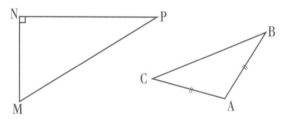

◆ Isosceles Triangles

• The **vertex angle** of an isosceles triangle is the angle whose sides are the two congruent sides of the triangle.

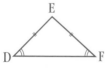

∠E is the vertex angle;
∠D and ∠F are the base angles;
$\overline{DF}$ is the base of ΔDEF;
$m\angle D = m\angle F$

NOTES
The equal number of curves in angles indicates
that they are equal in measurements.

- The base of an isosceles triangle is the side that
  does not have the same length as the other two
  sides, unless the triangle is equilateral. The
  base is not necessarily the side on the bottom
  of the triangle.
- The **base angles** of an isosceles triangle are the
  angles that have the base of the triangle as one
  of their sides.
- The base angles of an isosceles triangle are
  always equal in measure.

◆ **Right Triangles**
  - The **hypotenuse** of a right triangle is opposite
    the right angle and is the longest side of the
    right triangle.
  - The other two sides of a right triangle are
    called **legs**.

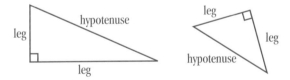

Pythagorean Theorem: The sum of the areas of the squares on the legs of a right triangle is equal to the area of the square on the hypotenuse; that is, the sum of the squares of the legs is equal to the square of the hypotenuse, or $a^2 + b^2 = c^2$ where a and b are the lengths of the legs and c is the length of the hypotenuse.

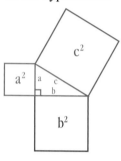

Notice the area of the small square is a^2, the medium square, b^2 and the largest square c^2. The areas of the 2 smaller squares equals the area of the biggest square; $\text{leg}^2 + \text{leg}^2 = \text{hypotenuse}^2$ or $a^2 + b^2 = c^2$

NOTES

Find the length of the hypotenuse of a right triangle when the lengths of the legs are 5 and 8. Find the length of a leg of a right triangle when the length of the hypotenuse is 12 and one of the legs is 9.

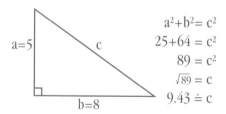

$$a^2 + b^2 = c^2$$
$$25 + 64 = c^2$$
$$89 = c^2$$
$$\sqrt{89} = c$$
$$9.43 \doteq c$$

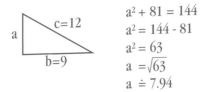

$$a^2 + 81 = 144$$
$$a^2 = 144 - 81$$
$$a^2 = 63$$
$$a = \sqrt{63}$$
$$a \doteq 7.94$$

NOTES

When looking for the hypotenuse, knowing both legs, **add** the squares and square root. When knowing the hypotenuse and looking for one leg **subtract** the squares and square root the value.

Theorem: If the square of the longest side of a triangle is equal to the sum of the squares of the other two sides, then the triangle is a right triangle.

Theorem: If the square of the longest side of the triangle is greater than the sum of the squares of the other two sides then the triangle is obtuse; if it is less than the sum of the squares of the other two sides then the triangle is acute.

45-45-90 Theorem: In a 45-45-90 triangle, the legs have equal lengths and the length of the hypotenuse is $\sqrt{2}$ times the length of one of the legs.

For example:
If the legs are 5 then the hypotenuse is 5 times $\sqrt{2}$ or $5\sqrt{2}$.

or

If the hypotenuse is 9 each leg is $9 \div \sqrt{2}$ or $\frac{9}{\sqrt{2}} \cdot \frac{\sqrt{2}}{\sqrt{2}} = \frac{9\sqrt{2}}{\sqrt{4}} = \frac{9\sqrt{2}}{2}$

30-60-90 Theorem: In a 30-60-90 triangle, the length of the shortest leg is $\frac{1}{2}$ of the length of the hypotenuse, and the length of the longer leg is $\sqrt{3}$ times the length of the shortest leg.

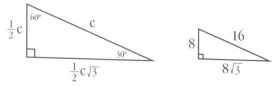

For example:
If the hypotenuse is 16 the shortest leg is 8 and the longest leg is $8\sqrt{3}$.

Theorem: The midpoint of the hypotenuse of a right triangle is equidistant (equal distances) from the three vertices of the triangle.

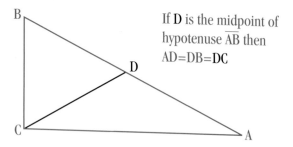

If **D** is the midpoint of hypotenuse $\overline{AB}$ then AD=DB=DC

Theorem: When an altitude is drawn to the hypotenuse of a right triangle,

◆ The two triangles that are formed are similar to each other and to the original right triangle.

$\overline{CD}$ is the altitude to hypotenuse $\overline{AB}$ so

$\triangle BDC \sim \triangle CDA \sim \triangle BCA$

and $\dfrac{BD}{CD} = \dfrac{CD}{DA}$ and $\dfrac{BA}{CB} = \dfrac{CB}{BD}$

and $\dfrac{BA}{CA} = \dfrac{CA}{AD}$

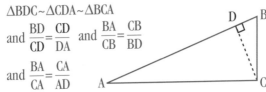

◆ The altitude is the geometric mean between the lengths of the two segments of the hypotenuse.

  • For example: In the right $\triangle$ above, if BD=4 and DA=15 then

$$\frac{4}{CD} = \frac{CD}{15}$$

$$CD^2 = 60$$

$$CD = \sqrt{60}$$

$$CD \doteq 7.75$$

- Each leg is the geometric mean between the hypotenuse and the length of the segment of the hypotenuse that is adjacent (touches) to the leg.

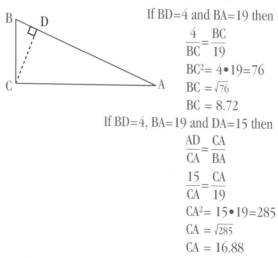

If BD=4 and BA=19 then

$$\frac{4}{BC}=\frac{BC}{19}$$

$$BC^2 = 4\bullet 19 = 76$$

$$BC = \sqrt{76}$$

$$BC = 8.72$$

If BD=4, BA=19 and DA=15 then

$$\frac{AD}{CA}=\frac{CA}{BA}$$

$$\frac{15}{CA}=\frac{CA}{19}$$

$$CA^2 = 15\bullet 19 = 285$$

$$CA = \sqrt{285}$$

$$CA = 16.88$$

Postulates & Theorems

› The sum of the 3 angle measurements of a triangle is 180°.

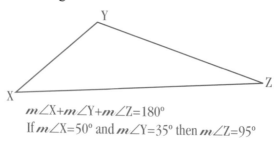

$m\angle X + m\angle Y + m\angle Z = 180°$

If $m\angle X = 50°$ and $m\angle Y = 35°$ then $m\angle Z = 95°$

- If two angle measurements of one triangle are equal to two angle measurements of another triangle then the measurements of the third angles are also equal.

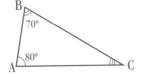

If $m\angle A=m\angle D=80°$ and $m\angle B=m\angle E=70°$ then $m\angle C=m\angle F=30°$

- Each angle of an equilateral triangle is 60°.

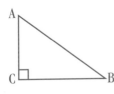

In $\triangle RST$, RS = ST = TR then $m\angle R = m\angle S = m\angle T$ and since the angles are equal and total 180°, each angle must equal 60° so $m\angle R = m\angle S = m\angle T = 60°$

- There can be no more than one right or obtuse angle in any one triangle.
- The acute angles of a right triangle are **complementary**.

$m\angle C+m\angle A+m\angle B=180$ and $m\angle C=90$ so, $90+m\angle A+m\angle B=180$ so, $m\angle A+m\angle B=90$ and $\angle A$ and $\angle B$ are complementary

- The measurement of an exterior angle of a triangle is equal to the sum of the measurements of the two remote (not having the same vertex as the exterior angle) interior angles of the triangle.

The exterior angle, $\angle 1$, is equal in measure to $m\angle X + m\angle Z$ because they all have different vertices, so

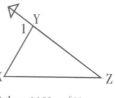

$m\angle 1 = m\angle X + m\angle Z$ and if $m\angle 1 = 110°$ and $m\angle X = 60°$ then $110° = 60° + m\angle Z$ and $50° = m\angle Z$ so $m\angle Y = 70°$

SSS Postulate: If three sides of one triangle are equal in length to three sides of another triangle, then the triangles are congruent (same shape and same size).

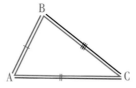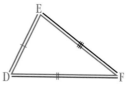

If $AB=DE$, $BC=EF$, and $AC=DF$ then $\triangle ABC \cong \triangle DEF$ therefore, $m\angle A = m\angle D$, $m\angle B = m\angle E$, and $m\angle C = m\angle F$
Note: In $\triangle ABC \cong \triangle DEF$, matching vertices are put in the same order; that is, $\triangle ABC \cong \triangle DEF$

SAS Postulate: If two sides and the included angle of one triangle are equal in measure to two sides and the included angle of another triangle, then the triangles are congruent.

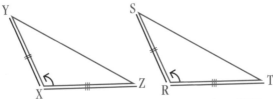

If XY = RS, XZ = RT, and $m\angle X = m\angle R$ then $\triangle XYZ \cong \triangle RST$. Note: $\angle X$ is between XY and XZ and $\angle R$ is between SR and RT. For example: If XY = SR = 8, XZ = RT = 15 and $m\angle X = m\angle R = 110°$ then $m\angle Y = m\angle S$, $m\angle Z = m\angle T$ and YZ = ST

ASA Postulate: If two angles and the included side of one triangle are equal in measure to two angles and the included side of another triangle, then the triangles are congruent.

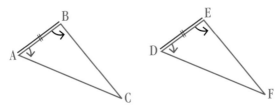

If $m\angle A = m\angle D$, $m\angle B = m\angle E$ and AB=DE, then $\triangle ABC \cong \triangle DEF$. Note: The side AB has the vertices of the $2\angle s$, A and B as endpoints and DE has the vertices of the $2\angle s$, D and E as endpoints

AA Similarity Postulate: If two angles of one triangle are equal in measure to two angles of another triangle, then the triangles are similar (same shape but not necessarily the same size).

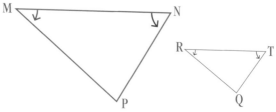

If $m\angle M = m\angle R$ and $m\angle N = m\angle T$ then $m\angle P$ must equal $m\angle Q$ because the sum of the angles of a $\Delta=180°$; if the angles of one Δ equal the angles of another Δ the shapes have to be the same, but the side lengths don't have to be equal; however, the sides must be proportional so $\frac{MN}{RT}=\frac{NP}{TQ}=\frac{PM}{QR}$

For example: If MN = 12, RT = 7 and NP = 10 then $\frac{MN}{RT}=\frac{NP}{TQ}$ so $\frac{12}{7}=\frac{10}{TQ}$ so TQ = 7•10÷12 = 5.83 and if RQ = 9 then $\frac{MN}{RT}=\frac{MP}{RQ}$ so $\frac{12}{7}=\frac{MP}{9}$ so MP = 12•9÷7=15.43

Theorem: If two sides of a triangle are equal in measure, then the angles opposite those sides are also equal in measure; and, if two angles of a triangle are equal in measure, then the sides opposite those angles are also equal in measure.

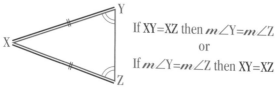

If XY=XZ then $m\angle Y=m\angle Z$
or
If $m\angle Y=m\angle Z$ then XY=XZ

Theorem: An equilateral triangle is also equiangular; and, an equiangular triangle is also equilateral.

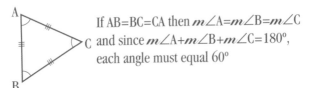

If AB=BC=CA then $m\angle A=m\angle B=m\angle C$ and since $m\angle A+m\angle B+m\angle C=180°$, each angle must equal 60°

Theorem: An equilateral triangle has three 60° angles.

Theorem: The bisector of the vertex angle of an isosceles triangle is the perpendicular bisector of the base of the triangle.

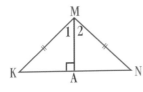

In isosceles $\triangle KMN$, if KM=MN then $\angle M$ is the vertex angle
If $m\angle 1=m\angle 2$, KA=AN and $\overline{MA}\perp\overline{KN}$

AAS Theorem: If two angles and a non-included side of one triangle are equal in measure to the two corresponding (matching if placed on top of the other shape) angles and non-included side of another triangle, then the triangles are congruent.

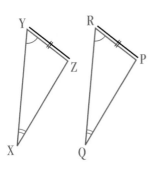

If $m\angle X=m\angle Q$, $m\angle Y=m\angle R$, and YZ=RP then $\triangle XYZ\cong$ $\triangle QRP$
Note: $\overline{YZ}$ is not located between $\angle X$ and $\angle Y$; $\overline{RP}$ is not located between $\angle R$ and $\angle Q$

HL Theorem: If the hypotenuse and one leg of a right triangle are equal in measure to the hypotenuse and the corresponding leg of another right triangle, then the two right triangles are congruent.

Remember, the hypotenuse is the side opposite the 90° angle and is the longest side of a right triangle.

SAS Inequality Theorem: If two sides of one triangle are equal in length to two sides of another triangle, but the included angle of one triangle is larger than the included angle of the other triangle, then the longer third side of the triangles is opposite the larger included angle of the triangles.

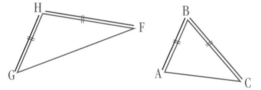

If GH=AB and HF=BC, but $m\angle H > m\angle B$ then GF>AC because GF is opposite the larger of the two angles H and B

SSS Inequality Theorem: If two sides of one triangle are equal in length to two sides of another triangle, but the third side of one triangle is longer than the third side of the other triangle, then the larger included angle (included between the two equal sides) is opposite the longer third side of the triangles.

This is the converse of the SAS Inequality Theorem above. It indicates that if GF>AC then

$m\angle H > m\angle B$

SSS Similarity Theorem: If the sides of one triangle are proportional to the corresponding sides of another triangle, then the triangles are similar.

SAS Similarity Theorem: If two sides of one triangle are proportional to two sides of another triangle and the included angles of each triangle are congruent, then the triangles are similar.

(See AA Postulate)

Triangle Proportionality Theorem: If a line is parallel to one side of a triangle and intersects the other two sides, then it divides those two sides proportionally, and creates 2 similar triangles.

(See AA Postulate)

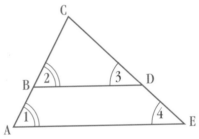

If BD $||$ AE then $m\angle 1 = m\angle 2$, $m\angle 4 = m\angle 3$, $\frac{BC}{AC} = \frac{DC}{EC}$, $\frac{BC}{BA} = \frac{DC}{DE}$, and $\triangle BCD \sim \triangle ACE$

For example: If BC=20, AC=28, and CD=22 then

$\frac{BC}{AC} = \frac{DC}{CE}$ so $\frac{20}{28} = \frac{22}{CE}$ so CE = $28 \bullet 22 \div 20 = 30.8$

Theorem: If a ray bisects an angle of a triangle, it divides the opposite side into segments proportional to the other two sides.

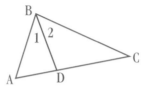

If $\overline{BD}$ bisects $\angle B$ then $m\angle 1 = m\angle 2$ and $\frac{AD}{DC} = \frac{AB}{BC}$. Ex: If AB=20, BC=24, and AD=13 then $\frac{13}{DC} = \frac{20}{24}$ so DC=24•13÷20=15.6 and AC=28.6

Theorem: The line segment that joins the midpoints of two sides of a triangle has two properties:
- it is parallel to the third side, and
- it is half the length of the third side.

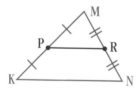

If P is the midpoint of $\overline{KM}$ and R is the midpoint of $\overline{MN}$ then $\overline{PR} \mathbin{\|} \overline{KN}$ and PR=$^1/_2$KN, so, if KN=30 then PR=15

Theorem: The 3 bisectors of the angles of a triangle intersect in one point which is equidistant from the 3 sides of the triangle.

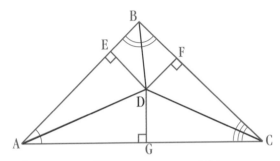

If $\overrightarrow{AD}$ bisects $\angle A$, $\overrightarrow{BD}$ bisects $\angle B$, and $\overrightarrow{CD}$ bisects $\angle C$ then they intersect in one point, D, and point D is equal distances from $\overline{AB}$, $\overline{BC}$ and $\overline{CA}$, so DE=DF=DG. Remember the distances from D to the sides must be $\perp$

Theorem: The perpendicular bisectors of the sides of a triangle intersect at one point which is equidistant from the 3 vertices of the triangle.

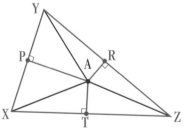

If $\overline{PA} \perp$ bisector of $\overline{XY}$, $\overline{RA} \perp$ bisector of $\overline{ZY}$, and $\overline{TA} \perp$ bisector of $\overline{XZ}$, then they all intersect at one point, A, which is equidistant from points X, Y, and Z, so XA=YA=ZA

For obtuse triangles, the point of intersection is in the exterior of the triangle; point D is equidistant from points X, Y, and Z, so XD=YD=ZD

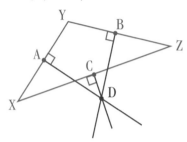

Theorem: The medians (line segments whose endpoints are one vertex of the triangle and the midpoint of the side that is opposite that vertex) of a triangle intersect in one point that is two thirds of the distance from each vertex to the midpoint of the opposite side.

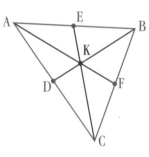

If points D, E and F are midpoints of the sides of △ABC then all medians intersect at one point, K; so, KB=²/₃DB, KC=²/₃EC, and KA=²/₃AF; it is also true that KB=2DK, KC=2EK, and KA=2FK

Theorem: If two sides of a triangle are unequal in length, then the angles opposite those sides are unequal and the larger angle is opposite the longer side; and, conversely, if two angles of a triangle are unequal, then the sides opposite those angles are unequal and the longer side is opposite the larger angle.

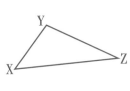

In scalene △XYZ, the largest angle is ∠Y and the longest side is opposite ∠Y, $\overline{XZ}$ The smallest angle is ∠Z so the shortest side is opposite ∠Z, $\overline{XY}$

Theorem: The sum of the lengths of any two sides of a triangle is greater than the length of the third side; and, the difference of the lengths of any two sides of a triangle is less than the length of the third side.

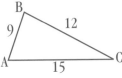

Note: 9+12>15, 15+12>9 and 9+15>12; also, 12−9<15, 15−12<9, and 15−9<12

■ Quadrilaterals

◆ Quadrilaterals are 4-sided polygons.

◆ Quadrilaterals have 2 diagonals and 4 vertices.

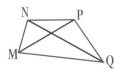

Quadrilateral MNPQ has sides $\overline{MN}$, $\overline{NP}$, $\overline{PQ}$ and $\overline{QM}$, with vertices M, N, P, and Q, and diagonals $\overline{NQ}$ and $\overline{MP}$

◆ **Special Quadrilaterals**

• **Trapezoids** are quadrilaterals with exactly one pair of parallel sides (called the bases), never more than one pair of parallel sides.

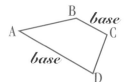

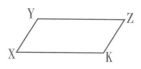

If $\overline{AD} \parallel \overline{BC}$ then quadrilateral ABCD is a trapezoid

If $\overline{XY} \parallel \overline{KZ}$ and $\overline{YZ} \parallel \overline{XK}$ then quadrilateral XYZK is not a trapezoid

› **Isosceles trapezoids** have nonparallel sides that are the same length and are called legs. The angles whose vertices are the endpoints of the same base of an isosceles trapezoid are called base angles, and they are equal in measure.

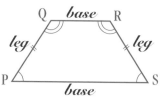

In trapezoid PQRS, if QP=RS then the trapezoid is isosceles; base angles of isosceles trapezoids are equal in measure so $m\angle P = m\angle S$ and $m\angle Q = m\angle R$

Theorem: The **median** of a trapezoid (the line segment whose endpoints are the midpoints of the 2 nonparallel sides of the trapezoid) is parallel to the bases, and has a length equal to half the sum of the lengths of the 2 bases (that is, equal to the average of the lengths of the 2 parallel sides, bases).

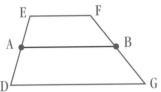

If A and B are midpoints of $\overline{DE}$ and $\overline{FG}$ then $\overline{AB} \parallel \overline{EF} \parallel \overline{DG}$ and AB = $1/2$(EF + DG) so if EF=10 and DG =18 then AB=$1/2$(10+18)=14

> The **area** of a trapezoid may be calculated by averaging the length of the bases and multiplying by the height (altitude—the line segment that forms 90° angles with the bases); thus the formula

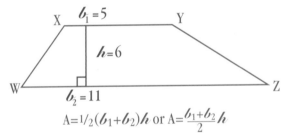

$A = 1/2(b_1 + b_2)h$ or $A = \dfrac{b_1 + b_2}{2}h$

NOTES
In this trapezoid, the parallel sides, that is the bases, are 5 and 11, so the average of the bases, $(5 + 11)/2$, is 8. Multiply 8 by the height of 6. Thus, the area is 48 square units.

• **Parallelograms** are quadrilaterals with 2 pairs of parallel sides.

Theorem: Opposite sides are parallel and equal in length.

Theorem: Opposite angles are equal in measure.

> All 4 interior angle measures total 360º.

> Consecutive interior angles (their vertices are endpoints for the same side of the parallelogram) are supplementary (measures total 180º).

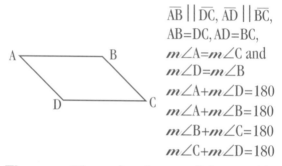

$\overline{AB} \,||\, \overline{DC}, \overline{AD} \,||\, \overline{BC},$
$AB=DC, AD=BC,$
$m\angle A = m\angle C$ and
$m\angle D = m\angle B$
$m\angle A + m\angle D = 180$
$m\angle A + m\angle B = 180$
$m\angle B + m\angle C = 180$
$m\angle C + m\angle D = 180$

Theorem: Diagonals of a parallelogram bisect each other.

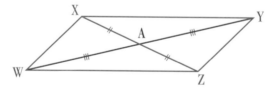

In $\square$WXYZ, diagonals $\overline{XZ}$ and $\overline{WY}$ bisect each other at A, so WA=AY and XA=AZ

Theorem: If one pair of opposite sides of a quadrilateral are equal in length and parallel, then the quadrilateral is a parallelogram.

Theorem: If both pairs of opposite sides of a quadrilateral are equal in length, then the quadrilateral is a parallelogram.

Theorem: If both pairs of opposite angles of a quadrilateral are equal, then the quadrilateral is a parallelogram.

Theorem: If the diagonals of a quadrilateral bisect each other, then the quadrilateral is a parallelogram.

› The **area of a parallelogram** can be calculated by multiplying the base and the height; that is, A=bh or A=hb. Note: Since opposite sides of a parallelogram are both parallel and equal in length, any side can be the base. The height (altitude) is any line segment that forms 90° angles with the base and whose endpoints are on the base and the opposite side of the parallelogram.

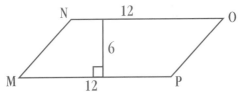

The base=12 and the height=6 so A=6•12=72 units^2

› **Special Parallelograms**

- **Rectangles** are parallelograms with 4 right angles (90° each).

Theorem: The diagonals of a rectangle are equal.

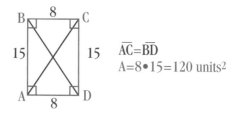

$\overline{AC}=\overline{BD}$

$A=8\bullet15=120$ units^2

The **area of a rectangle** is calculated by multiplying any 2 consecutive sides (sides that share a common endpoint). Since all angles are 90° any 2 consecutive sides are the base (length) and the height (width or altitude) of the rectangle; thus, A = bh or A = hb or A = lw.

- **Rhombuses (or Rhombi)** are parallelograms with 4 sides equal in length. The 4 interior angles can have any measures, but opposite angles have equal measures and all 4 angle measures total 360°.

Theorem: The diagonals of a rhombus are perpendicular.

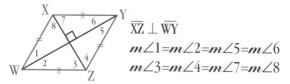

$\overline{XZ} \perp \overline{WY}$

$m\angle1=m\angle2=m\angle5=m\angle6$

$m\angle3=m\angle4=m\angle7=m\angle8$

Theorem: Each diagonal of a rhombus bisects the pair of opposite angles whose vertices are the endpoints of the diagonal.

- **Squares** have 4 equal sides and 4 equal angles (each 90º); therefore, every square is both a rectangle and a rhombus. Each square has 4 right angles just as all rectangles do, and each square has 4 equal sides just as all rhombi do. The diagonals of a square are equal in length, bisect each other, are perpendicular to each other, and bisect the interior angles of the square.

AB=BC=CD=DA
BD=AC
BD ⊥ AC
AE=ED=EC=EB

NOTES
The Venn diagram below indicates the relationships of squares to other quadrilaterals.

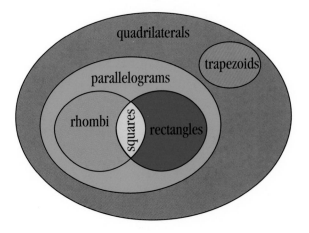

Circles

Defined Terms

■ A **circle** is the set of points in a plane that are the same distance from one point in the plane, which is called the **center** of the circle. The center of the circle lies in the interior of the circle and is not a point on the circle. ⊙ means circle.

■ The **radius** is the distance that each point on the circle is from the center of the circle; or, a radius is a line segment whose endpoints are the center of the circle and a point on the circle.

■ A **chord** is a line segment whose endpoints are 2 points on the circle.

■ A **diameter** is a chord that contains the center of the circle; or, a diameter is the length of the chord that contains the center of the circle (the distance from one point on the circle to another point on the circle, going through the center of the circle).

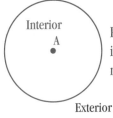

Point A is the center of ⊙A and is in the interior of the circle, not on the circle

■ A **secant** is a line that intersects a circle in two points.

■ A **tangent** is a line that is coplanar with a circle and intersects the circle in one point only. That point is called the **point of tangency.**

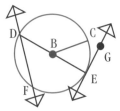

$\overline{BC}$ is the radius;

$\overline{DE}$ is a chord and a diameter;

$\overline{DF}$ is a chord;

$\overleftrightarrow{EG}$ is a tangent with E as the point of tangency;

$\overrightarrow{DF}$ is a secant

◆ A **common tangent** is a line that is tangent to 2 coplanar circles.

　• **Common internal tangents** intersect between the two circles.

　• Common external tangents do not intersect between the circles.

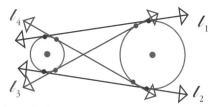

ℓ_1 and ℓ_2 are common external tangents.

ℓ_3 and ℓ_4 are common internal tangents.

• Two circles are tangent to each other when they are coplanar and share the same tangent line at the same point of tangency. They may be externally tangent or internally tangent.

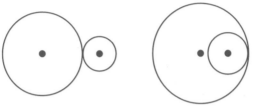

Externally tangent circles   Internally tangent circles

Equal circles are circles that have equal length radii (plural of radius).

Concentric circles are circles that lie in the same plane and have the same center.

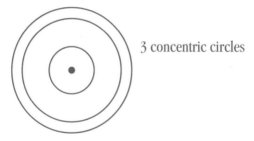

3 concentric circles

■ An **inscribed polygon** is a polygon whose vertices are points on the circle. In this same situation the circle is said to be **circumscribed about the polygon.**

Polygon ABCDE is inscribed in ⊙M; ⊙M is circumscribed about polygon ABCDE

Polygon WXYZ is not inscribed in ⊙P because vertex Z is not on ⊙P

■ **A circumscribed polygon** is a polygon whose sides are segments of tangents to the circle; i.e., the sides of the polygon each contain exactly one point on the circle. In this same situation the circle is said to be **inscribed in the polygon**.

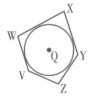

Pentagon VWXYZ is circumscribed about ⊙Q because each side is tangent to ⊙Q; ⊙Q is inscribed in pentagon VWXYZ

Polygon ABCD is NOT circumscribed about ⊙R because side $\overline{AD}$ is NOT tangent to ⊙R; ⊙R is NOT inscribed in polygon ABCD

■ An **arc** is part of a circle.
   ◆ A **semicircle** is an arc whose endpoints are the endpoints of a diameter. Three points must be used to name a semicircle.
   ◆ A **minor** arc is an arc whose length is less than the length of the semicircle. Only two points may be used to name a minor arc.
   ◆ A **major** arc is an arc whose length is more than the length of the semicircle. Three points must be used to name a major arc.

 Arc ABC or $\overarc{ABC}$ is a semicircle because chord $\overline{AC}$ is a diameter of $\odot P$; $\overarc{ADC}$ is also a semicircle; $\overarc{CD}$ and $\overarc{AD}$ are minor arcs; $\overarc{AD}=\overarc{DA}$ and $\overarc{CD}=\overarc{DC}$
$\overarc{DAB}, \overarc{BAD}, \overarc{BAC}, \overarc{BCA}, \overarc{DAC}, \overarc{DCA}$ are major arcs;
$\overarc{DAC}=\overarc{DBC}=\overarc{CBD}=\overarc{CAD}$; $\overarc{BAC}=\overarc{BDC}=\overarc{CAB}=\overarc{CDB}$

■ A **central angle** of a circle is an angle whose vertex is the center of the circle.
■ An **inscribed angle** is an angle whose vertex is on a circle and whose sides contain chords of the circle.

 $\angle APB$, $\angle BPC$, and $\angle APC$ are central angles because the vertex, P, is the center of the circle; $\angle FDE$ is an inscribed angle because the vertex, D, is on the circle

■ Theorems

◆ If a line is tangent to a circle, then the line is perpendicular to the radius whose endpoint is the point of tangency.

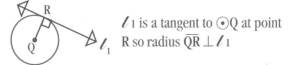

ℓ1 is a tangent to ⊙Q at point R so radius $\overline{QR} \perp \ell$1

◆ If a line in the plane of a circle is perpendicular to a radius at its outer endpoint, then the line is tangent to the circle.

◆ If two tangents intersect then the line segments whose endpoints are the point of intersection and the two points of tangency are equal in length; or, line segments drawn from a coplanar exterior point of a circle to points of tangency on the circle are equal in length.

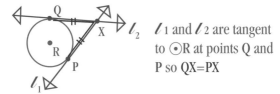

ℓ1 and ℓ2 are tangent to ⊙R at points Q and P so QX=PX

◆ The measure of a minor arc is equal to the measure of its central angle.

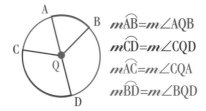

$m\widehat{AB}=m\angle AQB$

$m\widehat{CD}=m\angle CQD$

$m\widehat{AC}=m\angle CQA$

$m\widehat{BD}=m\angle BQD$

◆ The measure of a semicircle is 180º.

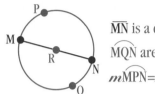

$\overline{MN}$ is a diameter so $\overparen{MPN}$ and $\overparen{MQN}$ are semicircles

$m\overparen{MPN}=m\overparen{MQN}=180º$

◆ The measure of a major arc is equal to 360º minus the measure of its corresponding minor arc.

$m\overparen{ACB}=360º-m\overparen{AB}$
If $m\overparen{AB}=90º$ then
$m\overparen{ACB}=270º$

◆ In the same circle or in equal circles, equal chords have equal arcs and equal arcs have equal chords.

If chords $\overline{AF}=\overline{BC}$ then $m\overparen{FA}=m\overparen{BC}$, but $ED\neq BC$ so $m\overparen{ED}\neq m\overparen{BC}$ and $m\overparen{ED}\neq m\overparen{AF}$

◆ A diameter or radius that is perpendicular to a chord bisects the chord and its arc.

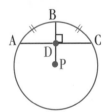

Radius $\overline{PB} \perp \overline{AC}$ so $\overline{AD}=\overline{DC}$ and $m\overparen{AB}=m\overparen{BC}$

◆ In the same circle or in equal circles, equal chords are the same distance from the center, and chords that are the same distance from the center are equal.

AD=BC if QE=QF because both chords are equal distances from the center, Q; remember, distance is always measured at 90° angles

◆ An inscribed angle is equal in measure to half of the measure of its intercepted arc (the arc which lies in the interior of the inscribed angle and whose endpoints are on the sides of the angle).

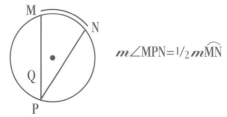
$m\angle MPN = 1/2\, m\overset{\frown}{MN}$

◆ If two inscribed angles intercept the same arc, then the angles are equal in measure.

∠ADB and ∠ACB are inscribed angles because the vertices, D and C, are on the circle
$m\angle ADB = 1/2\, m\overset{\frown}{AB}$ and $m\angle ACB = 1/2\, m\overset{\frown}{AB}$ so $m\angle ADB = m\angle ACB$

◆ If a quadrilateral is inscribed in a circle, then opposite angles are supplementary.

◆ An angle inscribed in a semicircle is always a right angle.

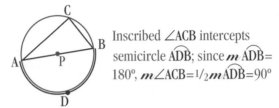

Inscribed $\angle ACB$ intercepts semicircle $\overset{\frown}{ADB}$; since $m\overset{\frown}{ADB}=180°$, $m\angle ACB=\frac{1}{2}m\overset{\frown}{ADB}=90°$

◆ The measure of an angle formed by a chord and a tangent is equal to half of the measure of its intercepted arc.

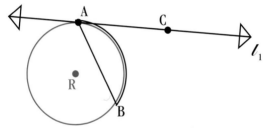

ℓ_1 is tangent to $\odot R$ at A so $m\angle BAC=\frac{1}{2}m\overset{\frown}{AB}$

If $m\overset{\frown}{AB}=110°$ then $m\angle BAC=55°$

◆ The measure of an angle formed by two chords intersecting inside a circle is equal to half the sum of the intercepted arcs.

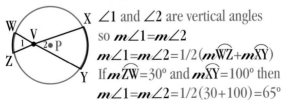

$\angle 1$ and $\angle 2$ are vertical angles so $m\angle 1=m\angle 2$

$m\angle 1=m\angle 2=\frac{1}{2}(m\overset{\frown}{WZ}+m\overset{\frown}{XY})$

If $m\overset{\frown}{ZW}=30°$ and $m\overset{\frown}{XY}=100°$ then $m\angle 1=m\angle 2=\frac{1}{2}(30+100)=65°$

◆ The measure of an angle formed by two secants, or two tangents, or a secant and a tangent, that intersect at a point outside of the circle, is equal to half the **difference** of the intercepted arcs.

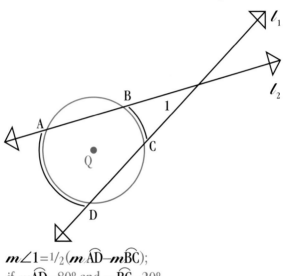

$m\angle 1 = \frac{1}{2}(m\stackrel{\frown}{AD} - m\stackrel{\frown}{BC})$;
if $m\stackrel{\frown}{AD} = 80°$ and $m\stackrel{\frown}{BC} = 20°$
then $m\angle 1 = \frac{1}{2}(80-20) = 30°$

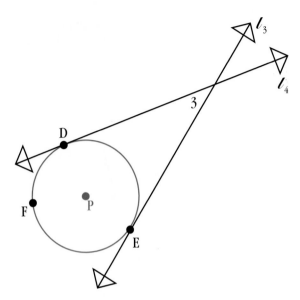

$\boldsymbol{\ell}_3$ and $\boldsymbol{\ell}_4$ are tangents to $\odot$P so
$m\angle 3 = \frac{1}{2}(m\overset{\frown}{DFE} - m\overset{\frown}{DE})$; if $m\overset{\frown}{DFE} = 210^\circ$ then
$m\overset{\frown}{DE} = 360^\circ - 210^\circ = 150^\circ$ and
$m\angle 3 = \frac{1}{2}(210-150) = 30^\circ$

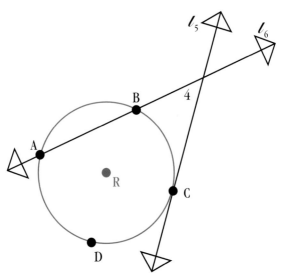

$\ell 5$ is a tangent to $\odot$R; $\ell 6$ is a secant;
$m\angle 4 = 1/2(m\overset{\frown}{ADC} - m\overset{\frown}{BC})$;
if $m\overset{\frown}{ADC}=170°$ and $m\overset{\frown}{BC}=84°$ then
$m\angle 4=1/2(170-84)=43°$

◆ When two chords intersect inside a circle, the
product of the segment lengths of one chord is
equal to the product of the segment lengths of
the other chord.

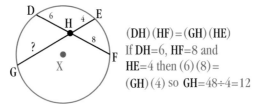

$(DH)(HF)=(GH)(HE)$
If **DH**=6, **HF**=8 and
HE=4 then $(6)(8)=$
$(GH)(4)$ so **GH**=48÷4=12

◆ When two secants are drawn to a circle from the same exterior point, the product of one secant and its external segment length equals the product of the other secant and its external segment length.

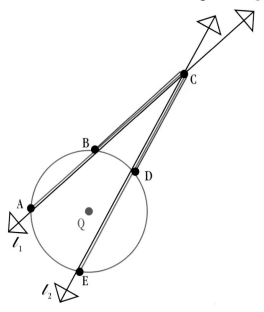

(BC)(AC)=(DC)(EC)
If AC=26, BC=14, and DC=10
then (14)(26)=(10)(EC) so
EC=14•26÷10=36.4 and since ED=EC-DC;
ED=36.4-10=26.4

♦ When a tangent and a secant are drawn to a circle from the same exterior point, the square of the length of the tangent segment is equal to the product of the secant and its external segment length.

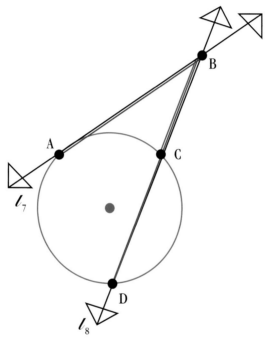

l_7 is tangent at point A. l_8 is a secant;
$(AB)(AB)=(DB)(CB)$ or $(AB)^2=(DB)(CB)$;
If AB=30, CB=20 then $(30)^2=(DB)(20)$ so
DB=$(30)^2 \div 20 = 900 \div 20 = 45$;
DC=DB-CB=45-20=25

Geometric Terms in Detail

Undefined Terms

■ **Point; notation**: Point A is labeled with a capital letter, A in this case.

■ **Line; notation**: Line KM is labeled either $\overleftrightarrow{KM}$ *or* $\overleftrightarrow{MK}$ or line *l*.

■ **Plane; notation**: Plane N is labeled either plane *n* or plane ABC if points A, B, and C are on plane *n*.

Defined Terms

General

■ **≅(congruent)**: Shapes are the same shape and size.

■ **~(similar)**: Shapes are the same shape, but can be different sizes.

■ **=(equal)**: Sets of points or numerical measurements are exactly the same.

■ **∪(union)**: Describes the result when all of the points are put together.

■ **∩(intersection)**: Describes the points where indicated shapes touch.

■ **Space**: The set of all points

Lines

■ **Collinear** points are on the same line.

■ **Non-collinear** points are not on the same line.

■ **Intersecting** lines have one and only one point in common.

■ **Perpendicular** lines intersect and form 90° angles at the intersection; ⊥

■ **Skew** lines are not in the same plane, never touch, and go in different directions.

■ **Transversal** lines intersect two or more coplanar lines at different points.

■ **Parallel** lines are coplanar (in the same plane), share no points in common, do not intersect, go in the same direction and never touch; ‖

Line Segments

■ The set of any 2 points on a line and all of the collinear points between them; $\overline{AB}$ where A and B are the endpoints of the line segment.

■ The **length** is the distance between the 2 endpoints; it is a numerical value; AB means the length of $\overline{AB}$.

■ The **midpoint** is a point exactly in the middle of the two endpoints of a line segment.

■ The **bisector** intersects a line segment at its midpoint.

■ The **perpendicular bisector** intersects a line segment at its midpoint and forms 90° angles at the intersection.

Rays

■ The set of collinear points going in one direction from one point (the endpoint of the ray) on a line; notation: $\overrightarrow{AB}$ where A is the endpoint; notice $\overrightarrow{AB} \neq \overrightarrow{BA}$ because they have different endpoints and contain different points on the line.

■ **Opposite rays** are collinear, share only a common endpoint and go in opposite directions.

Angles

■ The union of two rays that share one and only one point, the endpoint of the rays.
   ◆ The **sides** of the angle are the rays and the **vertex** is the endpoint of the rays.
   ◆ The **interior** is all the points between the two sides of the angle.
   ◆ ∠ABC where B is the vertex or simply ∠B if there is only one angle with vertex B.

■ **Overlapping angles** share some common interior points.

■ An **acute angle** measures less than 90°.

■ An **obtuse angle** measures more than 90°.

■ A **right angle** measures exactly 90°; it is indicated on diagrams by drawing a square in the corner by the vertex of the angle.

■ A **straight angle** measures exactly 180°.

■ **Complementary angles** are two angles whose measures total 90°.

■ **Supplementary angles** are two angles whose measures total 180°.

■ **Vertical angles** are two angles that share only a common vertex and whose sides form lines.

■ **Adjacent angles** are two angles that share exactly one vertex and one side, but no common interior points; i.e., they do not overlap.

■ An **angle bisector** is a ray or a line that contains the vertex of the angle, is in the interior, and separates the angle into two adjacent angles with equal measures.

Transversal Line Angles

■ **Interior angles** are formed with the rays from the 2 lines and the transversal such that the interior regions of the angles are located between the 2 lines.

■ **Alternate interior angles** are interior angles with different vertexes and interior regions on opposite sides of the transversal.

■ **Same side interior angles** are interior angles with different vertexes and interior regions on the same side of the transversal.

■ **Exterior angles** are formed with rays from the 2 lines and the transversal such that the interior regions of the angles are not between the 2 lines.

■ **Alternate exterior angles** are exterior angles with different vertexes and interior regions on opposite sides of the transversal.

■ **Corresponding angles** have different vertexes; their interior regions are on the same side of the transversal and in the same positions relative to the lines and the transversal; one of the pair of corresponding angles is an interior angle and the other is an exterior angle.

Polygons

■ Polygons are planar (flat), closed shapes that are formed by line segments that intersect only at their endpoints.

NOTES
Polygons are named by listing the endpoints of the line segments in order, going either clockwise or counterclockwise, starting at any one of the endpoints.

◆ The **sides** are line segments.
◆ The **interior** is all of the points enclosed by the sides.
◆ The **exterior** is all of the points on the plane of the polygon, but neither on the sides nor in the interior.

◆ The **vertices** (or vertexes) are the endpoints of the line segments.

◆ Include all the points on the sides (line segments) and the vertices.

◆ The **interior angles** of a polygon have the same vertices as the vertices of the polygon, have sides that contain the sides of the polygon, and have interior regions that contain the interior of the polygon—every polygon has as many interior angles as it has vertices.

◆ **Consecutive interior angles** have vertices that are endpoints of the same side of the polygon.

◆ The **exterior angles** are formed when the sides of the polygon are extended; each has a vertex and one side that are also a vertex and contain one side of the polygon; the second side of the exterior angle is the extension of the other polygon side containing the angle vertex; the interior of the exterior angle is part of the exterior region of the polygon; exterior angles are supplements of their adjacent interior angles.

◆ **Diagonals** of a polygon are line segments with endpoints that are vertices of the polygon, but the diagonals are not sides of the polygon.

■ **Concave** polygons have at least one interior angle measuring more than 180°.

■ **Convex** polygons have no interior angles more than 180° and all interior angles each measure less than 180°.

■ **Regular** polygons have all side lengths equal and all interior angle measures equal.

■ Classifications of Polygons

◆ Classified by the number of sides; equal to the number of vertices

◆ The side lengths and angle measures are not necessarily equal unless the word "regular" is also used to name the polygon.

◆ *Categories*
- Triangles have 3 sides
- Quadrilaterals have 4 sides
- Pentagons have 5 sides
- Hexagons have 6 sides
- Heptagons have 7 sides
- Octagons have 8 sides
- Nonagons have 9 sides
- Decagons have 10 sides
- *n*-gons have *n* sides

■ Special Polygons

◆ **Triangles**
- Polygons with 3 sides and 3 vertices; the symbol for a triangle is △; triangle ABC is written △***ABC***.
- An **altitude** (height) is a line segment with a vertex of the triangle as one endpoint and the point on the line containing the opposite side of the triangle where the altitude is perpendicular to that line; every triangle has 3 altitudes.
- A **base** is a side of the triangle on the line perpendicular to an altitude; every triangle has 3 bases.
- Formula for **area** $A = \frac{1}{2}ab$ or $A = \frac{1}{2}hb$

  where **a**=altitude, **b**=base or
  where **h**=height (altitude), **b**=base

NOTES
Triangles are classified in 2 ways, by side lengths and by angle measurements.

◆ **When classified by side lengths:**
   • **Scalene** have no side lengths =.
   • **Isosceles** have at least 2 side lengths equal.
   • **Equilateral** have all 3 side lengths equal.
     Note: It is also an isosceles triangle.

◆ **When classified by angle measurements:**
   • **Obtuse** have exactly one angle measurement more than 90°.
   • **Right** have exactly one angle measurement equal to 90°.
   • **Acute** have all 3 angles less than 90°.
     Note: that if all 3 angles are equal, then the triangle is called equiangular.

   • *Isosceles Triangles*
     › The vertex angle has sides containing the two congruent sides of the triangle.
     › The base is the side with a different length than the other two sides; not necessarily the side on the bottom of the triangle.
     › The base angles of an isosceles triangle have the base contained in one of their sides; they are always equal in measure.

   • *Right Triangles*
     › The hypotenuse is opposite the right angle and is the longest side.
     › The legs are the 2 sides that are not the hypotenuse; the line segments contained in the sides of the right angle.

◆ **Quadrilaterals**
  • 4-sided polygons
  • Have 2 diagonals and 4 vertices
  • **Trapezoids** have exactly one pair of parallel sides; there is never more than one pair of parallel sides.
    › Parallel sides: **bases**
    › Non-parallel sides: **legs**
    › The 2 angles with vertices that are the endpoints of the same base are called **base angles.**
    › **Isosceles trapezoids** have legs that are the same length.
  • **Parallelograms** have 2 pairs of parallel sides.
    › Rectangles have 4 right angles.
    › Rhombuses (sing. rhombus) have 4 sides equal in length.
    › Squares have 4 equal sides and 4 equal angles; therefore, every square is both a rectangle and a rhombus.

Circles

■ The set of points in a plane equidistant from the **center** of the circle, which lies in the interior of the circle and is not a point on the circle; 360°

■ A **radius** is a line segment whose endpoints are the center of the circle and any point on the circle; the length of a radius is the distance of each point from the center

■ A **chord** is a line segment whose endpoints are 2 points on the circle

■ A **diameter** is a chord that contains the center of the circle; the length of a diameter is the distance from one point to another on the circle, going through the center

■ A **secant** is a line intersecting a circle in two points

■ A **tangent** is a line that is co-planar with a circle and intersects it at one point only, called the point of tangency

  ◆ A **common tangent** is a line that is tangent to 2 co-planar circles

    • **Common internal tangents** intersect between the two circles

    • **Common external tangents** do not intersect between the circles

  ◆ Two circles are tangent when they are co-planar and share the same tangent line at the same point of tangency; they may be externally or internally tangent

■ **Equal circles** have equal length radii

■ **Concentric circles** lie in the same plane and have the same center

■ An **inscribed polygon** has vertices that are points on the circle; in this same situation the circle is circumscribed about the polygon

■ A **circumscribed polygon** has sides that are segments of tangents to the circle; i.e., the sides of the polygon each contain exactly one point on the circle; in this same situation, the circle is inscribed in the polygon

■ An **arc** is part of a circle

  ◆ A **semicircle** is an arc whose endpoints are the endpoints of a diameter; 180°; exactly three points must be used to name a semicircle; notation: $\overset{\frown}{ABC}$ where A and C are the endpoints of the diameter

  ◆ A **minor arc** length is less than the length of the semicircle; only two points may be used to name a minor arc; notation: $\overset{\frown}{DE}$ where D and E are the endpoints of the arc

◆ A **major arc** length is more than the length of the semicircle; exactly three points are used to name a major arc; notation: $\overgroup{FGH}$ where F and H are the endpoints of the arc

■ A **central angle** vertex is the center of the circle with sides that contain radii of the circle

■ An **inscribed angle** vertex is on a circle with sides that contain chords of the circle

Theorems & Relationships

Lines & Line Segments

■ Through a point not on a line, exactly one perpendicular can be drawn to the line.

■ The shortest distance from any point to a line or to a plane is the perpendicular distance.

■ Through a point not on a line, exactly one parallel can be drawn to the line.

■ Parallel lines are everywhere the same distance apart.

■ If three or more parallel lines cut off equal segments on one transversal, then they cut off equal segments on every transversal they share.

■ A line and a plane are parallel if they do not touch or intersect.

■ Two or more planes are parallel if they do not touch or intersect.

■ If two parallel planes are both intersected by a third plane, then the lines of intersection are parallel.

■ If a point lies on the perpendicular bisector of a line segment, then the point is equidistant (equal distances) from the endpoints of the line segment.

■ If a point is equidistant from the endpoints of a line segment, then the point lies on the perpendicular bisector of the line segment.

■ To trisect a line segment, separate it into three other congruent (equal in length) line segments such that the sum of the lengths of the three segments is equal to the length of the original line segment.

Angles

■ Angles are measured using a protractor and degree measurements: There are 360° in a circle; placing the center of a protractor at the vertex of an angle and counting the degree measure is like putting the vertex of the angle at the center of a circle and comparing the angle measure to the degrees of the circle.

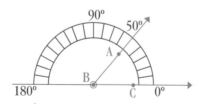

The measure of ∠ABC=m, ∠ABC=50°

NOTE: m∠ABC means the measure of the angle in degrees

■ If two angles are complements of the same angle then they are equal in measure (congruent).

■ If two angles are complements of congruent angles then they are congruent.

■ If two angles are supplements of the same angle then they are congruent.

■ If two angles are supplements of congruent angles then they are congruent.

■ Vertical angles are congruent and have equal measures.

■ If a point lies on the bisector of an angle, then the point is **equidistant** (equal distances) from the sides of the angle.

NOTES
Distance from a point to a line is always the length of the perpendicular line segment that has the point as one endpoint and a point on the line as the other.

■ If a point is equidistant from the sides of an angle, then the point lies on the bisector of the angle.

■ When an angle is trisected by rays or lines that contain the vertex of the angle and separate the angle into three adjacent angles (in pairs), all have equal measures.

Rays

■ If two rays do not intersect, then the union of the rays is simply all of the points on both rays.

■ If two rays intersect in one and only one point, but not at the endpoint, then the union is all of the points on both rays; the intersection is that one point where they touch.

■ If two rays intersect in one and only one point, the endpoint, then the union is an angle; the intersection is the endpoint.

$$\overrightarrow{AB} \cup \overrightarrow{AC} = \measuredangle CAB$$
$$\overrightarrow{AB} \cap \overrightarrow{AC} = A$$

■ If two rays intersect in more than one point then the union is a line; the intersection is a line segment.

$$\overrightarrow{AB} \cup \overrightarrow{BA} = \overleftrightarrow{AB}$$
$$\overrightarrow{AB} \cap \overrightarrow{BA} = \overline{AB}$$

Transversal Line Angles

■ If lines are parallel, then the alternate interior angles of a transversal are congruent.

■ If the alternate interior angles of a transversal are congruent, then the lines are parallel.

■ If lines are parallel, then the same side interior angles of a transversal are supplementary.

■ If the same side interior angles of a transversal are supplementary, then the lines are parallel.

■ If lines are parallel, then the corresponding angles of a transversal are congruent.

■ If the corresponding angles of a transversal are congruent then the lines are parallel.

■ If lines are parallel, then the alternate exterior angles of a transversal are congruent.

■ If the alternate exterior angles of a transversal are congruent, then the lines are parallel.

■ If a transversal is perpendicular to one of two parallel lines, then it is also perpendicular to the other.

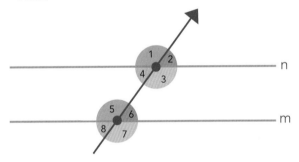

alternate interiors ∡: 4–6; 5–3

same side interiors ∡: 4–5; 3–6

correspondings ∡: 1–5; 4–8; 3–7; 2–6

alternate exterior ∡: 1–7; 2–8

Polygons

■ The sum of the measures of the interior angles of a convex polygon with n sides is $(n-2)180$ degrees.

> ### NOTES
> To find the measure of each interior angle of a regular polygon, find the sum of all of the interior angles and divide by the number of interior angles, thus the formula $\frac{(n-2)180}{n}$.

■ The sum of the measures of the exterior angles of any convex polygon, using one exterior angle at each vertex, is $360°$.

■ **Triangles**
   ◆ The 3-angle total measurement = $180°$.
   ◆ If two angle measurements of one triangle = two angle measurements of another triangle then the measurements of the third angles are also =.
   ◆ Each angle of an **equilateral triangle** is $60°$.
   ◆ There can be no more than one right or obtuse angle in any one triangle.
   ◆ The acute angles of a right triangle are complementary.
   ◆ The measurement of an exterior angle = the sum of the measurements of the two remote (not having the same vertex as the exterior angle) interior angles.
   ◆ If two sides of a triangle are equal, then the angles opposite to those sides are also equal; and, if two angles are equal, then the sides opposite those angles are also equal.

SAS Inequality Theorem: If two sides of one triangle are equal in length to two sides of another, but the included angle of one triangle is larger than the included angle of the other triangle, then the longer third side of the triangles is opposite the larger included angle of the triangles.

SSS Inequality Theorem: If two sides of one triangle are equal to two sides of another, but the third side of one is longer than the third side of the other, then the larger included angle (included between the two equal sides) is opposite to the longer third side of the triangles.

Triangle Proportionality Theorem: If a line is parallel to one side and intersects the other two sides, then it divides those two sides proportionally, and creates 2 similar triangles.

◆ If a *ray bisects* an angle of a triangle, it divides the opposite side into segments proportional to the other two sides.

◆ The line segment that joins the midpoints of two sides of a triangle has two properties:
   • It is *parallel* to the third side, and
   • It is *half the length* of the third side.

◆ The 3 bisectors of the angles of a triangle intersect in one point, which is equidistant from the 3 sides.

◆ The *perpendicular bisectors* of the sides of a triangle intersect in one point, equidistant from the 3 vertices.

◆ The medians (line segments whose endpoints are one vertex of the triangle and the midpoint of the side opposite that vertex) of a triangle intersect in one point two thirds of the distance from each vertex to the midpoint of the opposite side.

◆ If two sides of a triangle are unequal in length, then the opposite angles are unequal and the larger angle is opposite to the longer side; and conversely, if two angles of a triangle are unequal, then the sides opposite those angles are unequal and the longer side is opposite the larger angle.

◆ The sum of the lengths of any two sides is greater than the length of the third side; the difference of the lengths of any two sides is less than the length of the third side.

◆ **Isosceles & Equilateral Triangles**
  • An equilateral triangle is also equiangular; and, an equiangular triangle is also equilateral.
  • An equilateral triangle has three 60-degree angles.
  • The bisector of the vertex angle of an isosceles triangle is the perpendicular bisector of the base of the triangle.

◆ **Right Triangles**
  Pythagorean Theorem: In a right triangle, $a^2+b^2=c^2$, where a and b are the lengths of the legs and c is the length of the hypotenuse:
  • If the square of the hypotenuse is equal to the sum of the squares of the other two sides, then the triangle is a right triangle.

- If the square of the longest side is greater than the sum of the squares of the other two sides, then it is an obtuse triangle; if it is less than the sum of the squares of the other two sides, then it is an acute triangle.

  45-45-90 Theorem: In a 45-45-90 triangle, the legs have equal lengths and the length of the hypotenuse is $\sqrt{2}$ times the length of one of the legs.

  30-60-90 Theorem: In a 30-60-90 triangle, the length of the shortest leg is 1/2 the length of the hypotenuse, and the length of the longer leg is $\sqrt{3}$ times the length of the shortest leg.

- The midpoint of the hypotenuse of a right triangle is equidistant from the three vertices.
- When an altitude is drawn to the hypotenuse of a right triangle:
  › The two triangles formed are similar to each other and to the original right triangle
  › The altitude is the geometric mean between the lengths of the two segments of the hypotenuse.
  › Each leg is the geometric mean between the hypotenuse and the length of the segment of the hypotenuse adjacent (touches) to the leg.

◆ **Congruent Triangles**

  SSS Postulate: If three sides of one triangle are congruent to three sides of another, then the triangles are congruent.

  SAS Postulate: If two sides and the included angle of one triangle are congruent to two sides and the included angle of another, then the triangles are congruent.

ASA Postulate: If two angles and the included side of one triangle are congruent to two angles and the included side of another, then the triangles are congruent.

AAS Theorem: If two angles and a non-included side of one triangle are congruent to the two corresponding angles and non-included side of another, then the triangles are congruent.

HL Theorem: If the hypotenuse and one leg of a right triangle are congruent to the hypotenuse and the corresponding leg of another, then the two right triangles are congruent.

◆ **Similar Triangles**

AA Similarity Postulate: If two angles of one triangle are congruent to two angles of another, then the triangles are similar (same shape but not necessarily the same size).

SSS Similarity Theorem: If the sides of one triangle are proportional to the corresponding sides of another, then the triangles are similar.

SAS Similarity Theorem: If two sides of one triangle are proportional to two sides of another and the included angles of each triangle are congruent, then the triangles are similar.

■ Quadrilaterals
◆ Trapezoids
- The median (the line segment whose endpoints are the midpoints of the 2 non-parallel sides) is parallel to the bases, and its length is equal to half the sum of the lengths of the 2 bases.
- The area may be calculated by averaging the length of the bases and multiplying by the height (altitude that is the length of the line segment that forms 90 degree angles with the bases); thus the formula:

$$A = \frac{b_1 + b_2}{2} h = \frac{1}{2}(b_1 + b_2)h$$ where the 2 bases are

b_1 and b_2 and the height is h.
- Two angles with vertices that are the endpoints of the same leg of a trapezoid are **supplementary**.
- All 4 interior angle measures of all trapezoids total 360°.
- **Isosceles Trapezoid**
  › The base angles are congruent (has congruent legs).
  › Opposite angles are supplementary.

◆ Parallelograms
- Opposite sides are parallel and congruent.
- Opposite angles are congruent.
- All 4 interior angles total 360°.
- Consecutive interior angles (their vertices are endpoints for the same side) are supplementary.
- Diagonals bisect each other.
- A quadrilateral is a parallelogram if:
  › One pair of opposite sides is congruent and parallel.

> Both pairs of opposite sides are congruent.
> Both pairs of opposite angles are congruent.
> The diagonals bisect each other.

• The **area** can be calculated by multiplying the base and the height; that is, $A=bh=hb$.

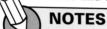

NOTES

Since opposite sides are both parallel and equal, any side can be the base: the height (altitude) is any line segment perpendicular to the base whose endpoints are on the base and the side opposite to the base.

• **Special Parallelograms**

> **Rectangles**
  - Parallelograms with 4 right angles
  - Diagonals are congruent and bisect each other.
  - The **area** equals lw or hb where l=length, w=width, h=height, and b=base.
  - If the 4 sides are all equal then the rectangle is more specifically called a square.

> **Rhombuses or Rhombi**
  - Parallelograms with 4 congruent sides.
  - Opposite angles are congruent.
  - All 4 angle measures total 360°.
  - Any 2 consecutive angles are supplementary.
  - If 4 interior angles each equal 90°; then the rhombus is more specifically called a square.
  - The diagonals are perpendicular bisectors of each other.
  - Each diagonal bisects the pair of opposite angles whose vertices are the endpoints of the diagonal.

> **Squares**
- 4 equal sides and 4 equal angles; every square is both a rectangle and a rhombus.
- The diagonals are congruent, bisect each other, are perpendicular to each other and bisect the interior angles.

NOTES
This Venn diagram indicates the relationships of quadrilaterals.

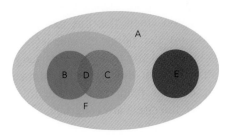

A = **Quadrilaterals**

B = **Rhombi**

C = **Rectangles**

D = **Squares**

E = **Trapezoids**

F = **Parallelograms**

Circles

■ If a line is **tangent** to a circle, then it is perpendicular to the radius whose endpoint is the point of tangency (the point where the tangent line intersects the circle).

l_1 is a tangent to ⊙ Q at point R so radius $\overline{QR} \perp l_1$

■ If two tangents to the same circle intersect in the exterior region, then the line segments whose endpoints are the point of intersection of the tangent lines and the two points of tangency are equal in length; or, line segments drawn from a coplanar exterior point of a circle to points of tangency on the circle are congruent.

■ If a line in the plane of a circle is perpendicular to a radius at its outer endpoint, then the line is tangent to the circle.

■ The measure of a **minor arc** is equal to the measure of its central angle.

■ The measure of a **semicircle** is 180°.

■ The measure of a **major arc** is equal to 360° minus the measure of its corresponding minor arc.

■ In the same circle or in equal circles, equal chords have equal arcs and equal arcs have equal chords.

■ A **diameter** perpendicular to a chord bisects the chord and its arc.

■ In the same circle or in equal circles, congruent chords are the same distance from the center, and chords the same distance from the center are congruent.

■ An **inscribed angle** is equal to half of its intercepted arc (the arc which lies in the interior of the inscribed angle and whose endpoints are on the sides of one angle).

$m\angle MPN = \frac{1}{2}m\overset{\frown}{MN}$

■ If two **inscribed angles** intercept the same arc, then the angles are congruent.

■ If a **quadrilateral** is inscribed in a circle, then opposite angles are supplementary.

■ An angle inscribed in a semicircle is always a right angle.

■ An angle formed by a **chord** and a **tangent** is equal to half of the measure of its intercepted arc.

■ An angle formed by two chords intersecting inside a circle is equal to half the sum of the intercepted arcs.

■ An angle formed by two secants, or two tangents, or a secant and a tangent, that intersect at a point outside of the circle is equal to half the difference of the intercepted arcs.

■ When two chords intersect inside a circle, the product of the segment lengths of one chord is equal to the product of the segment lengths of the other chord.

■ When two secant line segments are drawn to a circle from the same exterior endpoint, the product of one secant and its external segment length is equal to the product of the other secant and its external segment length.

■ When a tangent and a secant line segment are drawn to a circle from the same exterior point, the square of the length of the tangent segment is equal to the product of the secant and its external segment length.

9

Review of Geometric Formulas

■ **Area:** The area, A, of a 2-dimensional shape is the number of square units that can be put in the region enclosed by the sides.

> **NOTES**
> Area is obtained through some combination of multiplying heights and bases, which always form 90° angles with each other, except in circles.

■ **Perimeter:** The perimeter, P, of a 2-dimensional shape is the sum of all side lengths.

■ **Volume:** The volume, V, of a 3-dimensional shape is the number of cubic units that can be put in the space enclosed by all the sides.

Square Area: $A = b^2$
If $b = 8$, then:
$A = 64$ square units

Rectangle Area: $A = hb$, or $A = lw$
If $h = 4$ and $b = 12$ then:
$A = (4)(12)$, $A = 48$ square units

Triangle Area: $A = \frac{1}{2} bh$
If $h = 8$ and $b = 12$ then:
$A = \frac{1}{2}(8)(12)$, $A = 48$ square units

117

Parallelogram Area: $A = hb$
If $h = 6$ and $b = 9$ then:
$A = (6)(9)$, $A = 54$ square units

Trapezoid Area: $A = \frac{1}{2}h(b_1 + b_2)$
If $h = 9$, $b_1 = 8$ and $b_2 = 12$ then:
$A = \frac{1}{2}(9)(8 + 12)$, $A = \frac{1}{2}(9)(20)$,
$A = 90$ square units

Circle Area: $A = \pi r^2$
$A = \pi r^2$; If $r = 5$ then:
$A = \pi 5^2 = (3.14)25 = 78.5$ square units

Circumference: $C = 2\pi r$
If $r = 5$ then:
$C = (2)(3.14)(5) = 10(3.14) = 31.4$ units

Pythagorean Theorem:
If a right triangle has hypotenuse c
and legs a and b, then: $c^2 = a^2 + b^2$

Rectangular Prism Volume: $V = lwh$
If $l = 12$, $w = 3$ and $h = 4$ then:
$V = (12)(3)(4)$, $V = 144$ cubic units

Cube Volume: $V = e^3$
Each edge length, e, is equal to the
other edge in a cube; if $e = 8$ then:
$V = (8)(8)(8)$, $V = 512$ cubic units

Cylinder Volume: $V = \pi r^2 h$
If radius $r = 9$ and $h = 8$ then:
$V = \pi (9)^2(8)$, $V = (3.14)(81)(8)$,
$V = 2034.72$ cubic units

Cone Volume: $V = \frac{1}{3}\pi r^2 h$

If $r = 6$ and $h = 8$ then:

$V = \frac{1}{3}\pi (6)^2 (8)$, $V = \frac{1}{3}(3.14)(36)(8)$,

$V = 301.44$ cubic units

Triangular Prism Volume: $V = $ **(area of triangle)** h

If has an area equal to $\frac{1}{2}(5)(12)$ then:

$V = 30h$ and if $h = 8$ then:

$V = (30)(8)$, $V = 240$ cubic units

Rectangular Pyramid Volume: $V = \frac{1}{3}$ **(area of rectangle)** h

If $l = 5$ and $w = 4$, the rectangle has an area of 20, then:

$V = \frac{1}{3}(20)h$ and if $h = 9$ then:

$V = \frac{1}{3}(20)(9)$, $V = 60$ cubic units

Sphere Volume: $V = \frac{4}{3}\pi r^3$

If radius $r = 5$ then:

$V = \frac{4}{3}(3.14)(5)^3$, $V = 523.3$ cubic units

10 Postulates

NOTES
Postulates are statements that have been accepted without formal proof.

- A line contains at least 2 points, and any 2 points locate exactly one line.

- Any 3 non-collinear points locate exactly one plane.

- A line and one point not on the line locate exactly one plane.

- Any 3 points locate at least one plane

- If 2 points of a line are in a plane, then the line is in the plane.

- If 2 points are in a plane, then the line containing the 2 points is also in the plane.

- If 2 planes intersect, then the intersection is a line.